Differential equations

Differential equations

Their solution using symmetries

HANS STEPHANI

Sektion Physik, Friedrich-Schiller-Universität, Jena, G. D. R.

Edited by

MALCOLM MACCALLUM

School of Mathematical Sciences, Queen Mary College, London

The right of the University of Cambridge to print and sell all manner of books was granted by Henry VIII in 1534. The University has printed and published continuously since 1584.

CAMBRIDGE UNIVERSITY PRESS

Cambridge

New York Port Chester

Melbourne Sydney

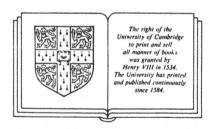

CAMBRIDGE UNIVERSITY PRESS
Cambridge, New York, Melbourne, Madrid, Cape Town, Singapore,
São Paulo, Delhi, Dubai, Tokyo, Mexico City

Cambridge University Press
The Edinburgh Building, Cambridge CB2 8RU, UK

Published in the United States of America by
Cambridge University Press, New York

www.cambridge.org
Information on this title: www.cambridge.org/9780521366892

First published 1989

A catalogue record for this publication is available from the British Library

Library of Congress Cataloguing in Publication Data

Stephani, Hans.
Differential equations: their solution using symmetries/Hans Stephani; edited
by Malcolm MacCallum.
p. cm.
Includes index.
ISBN 0-521-35531-1. – ISBN 0-521-36689-5 (pbk.)
1. Differential equations – Numerical solutions. 2. Symmetry.
I. MacCallum, M. A. H. II. Title.
QA371.S8473 1989
515.3′5 – dc19 88-36744
CIP

ISBN 978-0-521-35531-1 Hardback
ISBN 978-0-521-36689-2 Paperback

Contents

Preface

Several years ago, when reading an old paper on a special class of solutions to Einstein's field equations, I became aware by pure chance of the existence of Lie's method of integrating ordinary differential equations when their symmetries are known. Since I had never heard of or read about this method, I consulted friends and textbooks to make sure that this was not only my personal fault and ignorance. Lie invented the theory of Lie groups when studying symmetries of differential equations. Lie groups became well known, but their original field of fruitful application remained hidden in the literature and either did not get into textbooks, or, if it did, only appeared in very mutilated versions. Hence many people who could have profited from this method simply were not aware of its existence. Even of the monographs then available, most contained only part of what I found when reading Lie's original papers (which fortunately enough were written in German). With the eagerness of a freshly baptized believer, I therefore decided to give a series of lectures on this subject, lectures that eventually developed into this book.

This book is intended to be an introduction. Almost nothing written in it is my own. But it is rather difficult to give proper credit to the origins of the various ideas and results – not only because of the gaps in my memory, but also because most of the material presented here is due to Lie, with only minor improvements and generalizations by later workers who rediscovered and used Lie's results.*

* Many remarks on the historical development of the subject can be found in Olver's textbook.

This book could not have been written without the help and advice given to me in many discussions, in particular by my colleagues Dietrich Kramer, Malcolm MacCallum, and Gernot Neugebauer, and could not have been printed without the innumerable amendments in style and grammar by which Malcolm MacCallum tried to make my English readable. I have to thank them all, and also Frau U. Kaschlik for the careful typing of the manuscript.

Jena/Crawinkel Hans Stephani

1

Introduction

Suppose you have to solve an ordinary differential equation of, say, second order, for example,

$$y'' = (x - y)y'^3, \qquad y' \equiv dy/dx, \quad \text{etc.,} \tag{1.1}$$

or

$$y'' = x^n y^2, \qquad n \neq 0. \tag{1.2}$$

How can you proceed? First, you will certainly check whether the differential equation belongs to a class you know how to treat, for preference to the class of linear differential equations, or whether it can be transformed into a member of such a class by simple transformations of the dependent and independent variables. If unsuccessful, you may look up a textbook or a collection of solutions. And if that does not help and the problem is of some importance, you will try more elaborate transformation methods and ad hoc *ansätze* to find the solution. But it sometimes may and will happen that you have to leave the problem unsolved, with the uneasy feeling that there might be a method you have overlooked or you are unaware of, and the question of whether the solution to your differential equation can be given in terms of elementary, simple functions will still be open. [For the differential equations (1.1) and (1.2), the answer to that question will be given in Sections 5.2 and 7.5.]

For a partial differential equation such as

$$u_{,xx}u_{,yy} - u_{,xy}^2 - 1 = 0, \tag{1.3}$$

the situation is similar, although the specific questions are different. Since really powerful methods to deal with partial differential equations of second order are easily available only for linear equations, you may first try to transform the given differential equation into a linear one [for the example (1.3) that can be done, see Exercise 21.4.6]. If unsuccessful, you may perhaps set out to find as many special solutions as possible by using standard techniques such as separation of variables (writing u as a product or a sum of functions of different variables) or reduction of variables (assuming that u depends on less independent variables). But again the question remains open whether these methods will work and whether there are other methods that may prove useful.

In both cases, for ordinary and for partial differential equations, the unmistakable answer to all these questions is that in the majority of the cases where exact solutions of a differential equation can be found, the underlying property is a symmetry of that equation. Moreover, there are practicable methods to find those symmetries (if they exist) and to use them in solving the differential equation.

This book is devoted to the study of symmetries of differential equations, with the emphasis on how to use symmetries to find solutions. The typical way to define and find a symmetry is to first characterize a class of admitted transformations of variables and then look for special transformations in this class that leave a given differential equation invariant. The more general the admissible transformations are, the more symmetries will exist – but the more difficult it will be to find them and to exploit them to give an integration procedure. I have therefore arranged the material according to the generality of the allowed transformations, beginning with Lie point transformations and symmetries and ending with generalized (dynamical and Lie–Bäcklund) symmetries.

Some aspects of the theory could have been treated simultaneously for ordinary and for partial differential equations. To make the book easier to read, the subject has been divided. Thus some paragraphs of the second part on partial differential equations are in essence slightly generalized repetitions of what has already been said in the first part in the context of ordinary differential equations. Although that does not imply that the second part is completely self-contained, one can start reading there if interested only in partial differential equations, at the price of meeting a comparatively concise presentation from the very beginning.

The book is written in an intuitive fashion; none of the proofs is really watertight in that the necessary assumptions are not clearly stated (e.g., most of the functions that appear are tacitly assumed to be differentiable or even analytic). The emphasis is always on presenting the main ideas and giving advice on how to use them in practice. Most of the material is standard and long known. No references are given, therefore, in the text, but a small guide to the existing literature is added as an appendix.

I

Ordinary differential equations

2

Point transformations and their generators

2.1 One-parameter groups of point transformations and their infinitesimal generators

As stated in the introduction, our main goal is to use symmetries of differential equations for their integration. To do this, we need a proper definition of a symmetry, and this in turn requires some knowledge of transformations and their generators.

When dealing with differential equations, one very often tries to simplify the equation by an appropriate change of variables, that is, by a transformation of the independent variable x and the dependent variable y,

$$\tilde{x} = \tilde{x}(x, y), \qquad \tilde{y} = \tilde{y}(x, y). \tag{2.1}$$

We call this a *point transformation* (as contrasted with, e.g., contact transformations, which will be discussed in Chapter 11); it maps points (x, y) into points $(\tilde{x}, \tilde{y})$.

In the context of symmetries we have to consider point transformations that depend on (at least) one arbitrary parameter ε,

$$\tilde{x} = \tilde{x}(x, y; \varepsilon), \qquad \tilde{y} = \tilde{y}(x, y; \varepsilon), \tag{2.2}$$

and that furthermore have the properties that they are invertible, that repeated applications yield a transformation of the same family, for example,

$$\tilde{\tilde{x}} = \tilde{\tilde{x}}(\tilde{x}, \tilde{y}; \tilde{\varepsilon}) = \tilde{\tilde{x}}(x, y; \tilde{\tilde{\varepsilon}}) \tag{2.3}$$

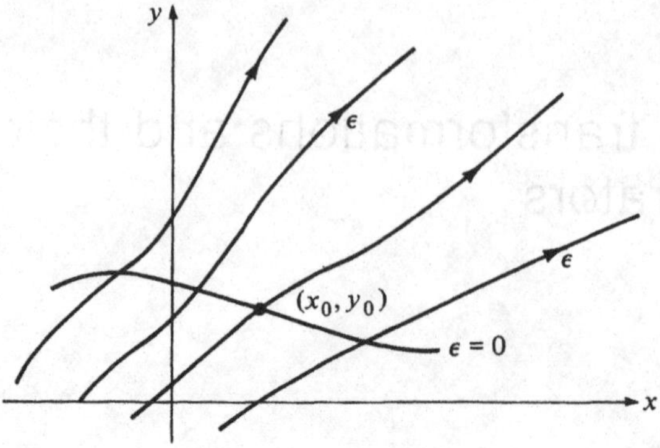

Figure 2.1. Action of a one-parameter group of transformations.

for some $\tilde{\tilde{\varepsilon}} = \tilde{\tilde{\varepsilon}}(\tilde{\varepsilon}, \varepsilon)$, and that the identity is contained for, say, $\varepsilon = 0$:

$$\tilde{x}(x, y; 0) = x, \qquad \tilde{y}(x, y; 0) = y. \tag{2.4}$$

In fact, these properties ensure that the transformations (2.2) form a *one-parameter group of point transformations.*

A simple example of a one-parameter group is given by the rotations

$$\tilde{x} = x \cos \varepsilon - y \sin \varepsilon, \qquad \tilde{y} = x \sin \varepsilon + y \cos \varepsilon. \tag{2.5}$$

On the other hand, the reflection

$$\tilde{x} = -x, \qquad \tilde{y} = -y \tag{2.6}$$

is a point transformation that, although useful, does not constitute a one-parameter group.

The one-parameter group (2.2) and its action can best be visualized as motion in an x–y plane. To do this, take (for $\varepsilon = 0$) an arbitrary point (x_0, y_0) in that plane. When the parameter ε varies, the images $(\tilde{x}_0, \tilde{y}_0)$ of (x_0, y_0) will move along some line. Repeating this for different initial points, one obtains the picture given in Figure 2.1, each curve representing points that can be transformed into each other under the action of the group. They are called the orbits of the group. This picture can also be interpreted in terms of the flow and the stream lines of some fluid.

The picture of Figure 2.1 suggests that a different representation of the transformation group given by equations (2.2) should be possible: the set of curves given in Figure 2.1 is completely characterized by the field of its tangent vectors $\mathbf{X}$, see Figure 2.2, and vice versa!

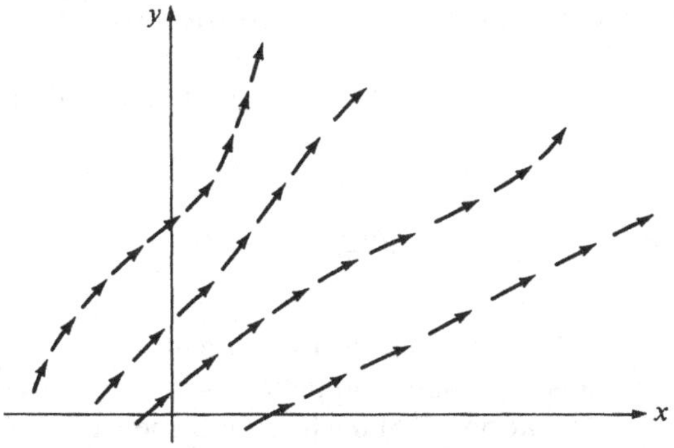

Figure 2.2. The field of tangent vectors **X** associated with orbits of a one-parameter group.

This idea can be given a concise form by considering infinitesimal transformations. We take an arbitrary point (x, y) and write

$$\tilde{x}(x, y; \varepsilon) = x + \varepsilon\xi(x, y) + \cdots = x + \varepsilon\mathbf{X}x + \cdots,$$
$$\tilde{y}(x, y; \varepsilon) = y + \varepsilon\eta(x, y) + \cdots = y + \varepsilon\mathbf{X}y + \cdots, \tag{2.7}$$

where the functions ξ and η are defined by

$$\xi(x, y) = \left.\frac{\partial\tilde{x}}{\partial\varepsilon}\right|_{\varepsilon=0}, \qquad \eta(x, y) = \left.\frac{\partial\tilde{y}}{\partial\varepsilon}\right|_{\varepsilon=0}, \tag{2.8}$$

and the operator **X** is given by

$$\mathbf{X} = \xi(x, y)\frac{\partial}{\partial x} + \eta(x, y)\frac{\partial}{\partial y}. \tag{2.9}$$

Obviously, the components of the tangent vector **X** are exactly ξ and η.

The operator **X** is called the *infinitesimal generator* of the transformation. "Generator" indicates that repeated application of the infinitesimal transformation will generate the finite transformation, which is a different way of expressing the fact that the integral curves of the vector field **X** are the group orbits: that is, by integrating

$$\frac{\partial\tilde{x}}{\partial\varepsilon} = \xi(\tilde{x}, \tilde{y}), \qquad \frac{\partial\tilde{y}}{\partial\varepsilon} = \eta(\tilde{x}, \tilde{y}) \tag{2.10}$$

with the initial values x, y at $\varepsilon = 0$, we will arrive at the finite transformation (2.2).

The infinitesimal generator uniquely determines the orbits of the group, but the orbits will give the generator only up to a constant factor: if we rescale the parameter ε by $\varepsilon = f(\hat{\varepsilon}), f(0) = 0, f'(0) \neq 0$, the definition (2.8) of ξ and η yields

$$\hat{\xi} = \frac{\partial \tilde{x}}{\partial \hat{\varepsilon}}\bigg|_{\hat{\varepsilon}=0} = \frac{\partial \tilde{x}}{\partial \varepsilon} f'(\hat{\varepsilon})\bigg|_{\varepsilon=0} = f'(0)\xi, \quad \hat{\eta} = f'(0)\eta \tag{2.11}$$

(the tangent vector $\mathbf{X}$ of the orbits has no fixed scale).

As an illustration of transformations and their generators, let us consider some examples. For the rotations (2.5) in the x–y plane we have

$$\frac{\partial \tilde{x}}{\partial \varepsilon}\bigg|_{\varepsilon=0} = -y, \qquad \frac{\partial \tilde{y}}{\partial \varepsilon}\bigg|_{\varepsilon=0} = x, \tag{2.12}$$

so that the corresponding generator is given by

$$\mathbf{X} = -y\frac{\partial}{\partial x} + x\frac{\partial}{\partial y}. \tag{2.13}$$

For a translation (shift of the origin of x) we obtain

$$\tilde{x} = x + \varepsilon, \qquad \tilde{y} = y, \qquad \mathbf{X} = \frac{\partial}{\partial x}. \tag{2.14}$$

The inverse problem is to find the finite transformation when the generator is given. If we have

$$\mathbf{X} = x\frac{\partial}{\partial x} + y\frac{\partial}{\partial y}, \tag{2.15}$$

which group corresponds to it? Of course we have to integrate (2.10), that is,

$$\frac{\partial \tilde{x}}{\partial \varepsilon} = \tilde{x}, \qquad \frac{\partial \tilde{y}}{\partial \varepsilon} = \tilde{y}. \tag{2.16}$$

The solution with initial values $\tilde{x}(0) = x$, $\tilde{y}(0) = y$ is obviously

$$\tilde{x} = e^{\varepsilon}x, \qquad \tilde{y} = e^{\varepsilon}y. \tag{2.17}$$

This is a (special) scaling – or similarity – transformation: all variables are multiplied by the same constant factor.

Looking back at the last few pages, the reader may wonder why we so strongly preferred the transformations that form a (Lie) group and the generators of those transformations. The reason is that although transformations (2.1) that are not members of one-parameter groups may and will occur in the context of symmetries of differential equations, it is only with the help of the generators **X** that we will be able to *find* (and use) symmetries. This is mainly due to the fact that although the transformations themselves can be very complicated, the generators are always *linear* operators.

2.2 Transformation laws and normal forms of generators

The generators **X** as given by (2.9) explicitly refer to the variables x and y. How do the components ξ and η change if we introduce new variables $u(x, y)$ and $v(x, y)$ instead of x and y?

We slightly generalize the question by considering generators in more than two variables,

$$\mathbf{X} = b^i(x^n)\frac{\partial}{\partial x^i}, \qquad i = 1, \ldots, N \tag{2.18}$$

(summation over the repeated index i). Performing a transformation

$$x^{i'} = x^{i'}(x^i), \qquad |\partial x^{i'}/\partial x^i| \neq 0, \tag{2.19}$$

gives because of

$$\frac{\partial}{\partial x^i} = \frac{\partial x^{i'}}{\partial x^i}\frac{\partial}{\partial x^{i'}} \tag{2.20}$$

the transformation law

$$\mathbf{X} = b^{i'}\frac{\partial}{\partial x^{i'}}, \qquad b^{i'} = \frac{\partial x^{i'}}{\partial x^i}b^i. \tag{2.21}$$

As is to be expected, the components b^i of the generator **X** transform as the (contravariant) components of a vector.

We can make use of this transformation law to write the generator **X** in a different form. Because of

$$\mathbf{X}x^n = b^i\frac{\partial}{\partial x^i}x^n = b^n, \qquad \mathbf{X}x^{n'} = b^{n'}, \tag{2.22}$$

this form is

$$\mathbf{X} = (\mathbf{X}x^i)\frac{\partial}{\partial x^i} = (\mathbf{X}x^{i'})\frac{\partial}{\partial x^{i'}}. \tag{2.23}$$

It clearly indicates how to calculate the components of **X** in the new coordinates $x^{i'}(x^i)$ if **X** is known in coordinates x^i: one simply has to apply **X** to the new coordinates.

If, for example, we want to express the generator (2.15) in coordinates

$$u = y/x, \qquad v = xy, \tag{2.24}$$

we obtain immediately from (2.23).

$$\mathbf{X}u = \left(x\frac{\partial}{\partial x} + y\frac{\partial}{\partial y}\right)u = 0, \qquad \mathbf{X}v = 2xy = 2v, \tag{2.25}$$

that is,

$$\mathbf{X} = 2v\frac{\partial}{\partial v}. \tag{2.26}$$

A second example is provided by the generator (2.13) of a rotation if we want to express it in polar coordinates $r = (x^2 + y^2)^{1/2}$, $\varphi = \arctan y/x$. The result is $\mathbf{X}r = 0$, $\mathbf{X}\varphi = 1$, that is,

$$\mathbf{X} = -y\frac{\partial}{\partial x} + x\frac{\partial}{\partial y} = \frac{\partial}{\partial \varphi}. \tag{2.27}$$

Obviously polar coordinates are better adjusted to describing rotations than Cartesian coordinates are. This (trivial) statement leads to the question: are there always coordinates that are maximally adjusted to a given one-parameter group of transformations? The answer is yes, there always exist coordinates such that – for an arbitrary number N of coordinates x^i – the generator (2.18) takes the simple form

$$\mathbf{X} = \frac{\partial}{\partial s}. \tag{2.28}$$

We call (2.28) the *normal form* of the generator **X**.

To prove the above assertion, we could refer to the theory of partial differential equations, which shows that the system of equations

$$\mathbf{X}s = b^i\frac{\partial s}{\partial x^i} = 1, \qquad \mathbf{X}x^{n'} = b^i\frac{\partial x^{n'}}{\partial x^i} = 0,$$

$$i = 1,\dots,N, \quad n' = 2,\dots,N, \tag{2.29}$$

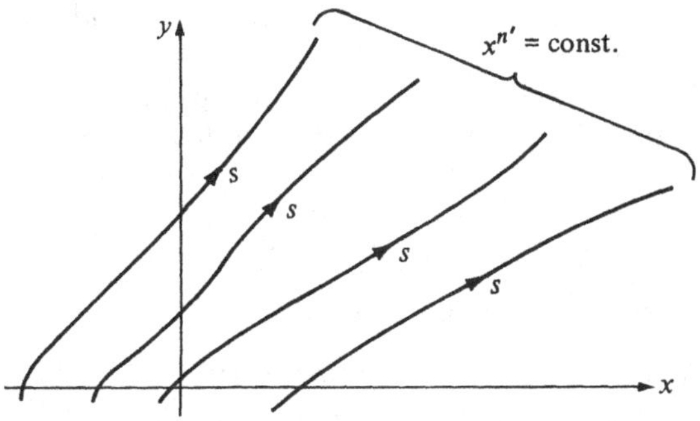

Figure 2.3. Transformation to coordinates $(s, x^{n'})$ for which $\mathbf{X} = \partial/\partial s$.

always has a nontrivial solution $\{s(x^i), x^{n'}(x^i)\}$. But it is more instructive to refer to Figure 2.3 and simply to state that if in an N-dimensional space (coordinates x^i) we have a congruence of one-dimensional group orbits covering the space smoothly and completely, then we can take these curves as coordinate lines $x^{n'} = \text{const}$ and take s as a parameter along these curves. In these coordinates the only non-zero component of the tangent vector $\mathbf{X}$ is in the s-direction, and by a suitable choice of the parameter s (different on each orbit!) we can arrange that this s-component equals unity.

To find the explicit transformation that brings a given generator into its normal form $\mathbf{X} = \partial/\partial s$ can be a difficult task. But in most applications that occur in the context of differential equations we shall be able to perform it. Comparing Figures 2.1 and 2.3, we see that it obviously amounts to finding the finite transformation and then converting this to obtain $\varepsilon = s(x, y)$, see Exercise 2.2.

2.3 Extensions of transformations and their generators

If we want to apply a point transformation (2.1) or (2.2) to a differential equation

$$H(x, y, y', y'', \ldots, y^{(n)}) = 0, \qquad y' \equiv \mathrm{d}y/\mathrm{d}x, \quad \text{etc.}, \tag{2.30}$$

we have to know how to transform the derivatives $y^{(n)}$, that is, how to extend (or to prolong) the point transformation to the derivatives. Of course, this is trivially done by defining

$$\tilde{y}' = \frac{\mathrm{d}\tilde{y}}{\mathrm{d}\tilde{x}} = \frac{\mathrm{d}\tilde{y}(x, y; \varepsilon)}{\mathrm{d}\tilde{x}(x, y; \varepsilon)} = \frac{y'(\partial\tilde{y}/\partial y) + (\partial\tilde{y}/\partial x)}{y'(\partial\tilde{x}/\partial y) + (\partial\tilde{x}/\partial x)} = \tilde{y}'(x, y, y'; \varepsilon),$$

$$\tilde{y}'' = \frac{\mathrm{d}\tilde{y}'}{\mathrm{d}\tilde{x}} = \tilde{y}''(x, y, y', y''; \varepsilon), \quad \text{etc.}; \tag{2.31}$$

that is, the transformed derivatives are the derivatives of, and with respect to, the transformed variables.

What we really need are the extensions (or prolongations) of the infinitesimal generators **X**. We obtain them as follows. We write – as in (2.7) –

$$
\begin{aligned}
\tilde{x} &= x + \varepsilon\xi(x, y) + \cdots = x + \varepsilon \mathbf{X} x + \cdots, \\
\tilde{y} &= y + \varepsilon\eta(x, y) + \cdots = y + \varepsilon \mathbf{X} y + \cdots, \\
\tilde{y}' &= y' + \varepsilon\eta'(x, y, y') + \cdots = y' + \varepsilon \mathbf{X} y' + \cdots, \\
&\vdots \\
\tilde{y}^{(n)} &= y^{(n)} + \varepsilon\eta^{(n)}(x, y, y', \ldots, y^{(n)}) + \cdots = y^{(n)} + \varepsilon \mathbf{X} y^{(n)} + \cdots,
\end{aligned}
\tag{2.32}
$$

where $\eta, \eta', \ldots, \eta^{(n)}$ are *defined* by

$$
\eta' = \left.\frac{\partial \tilde{y}'}{\partial \varepsilon}\right|_{\varepsilon=0}, \ldots, \eta^{(n)} = \left.\frac{\partial \tilde{y}^{(n)}}{\partial \varepsilon}\right|_{\varepsilon=0}. \tag{2.33}
$$

Inserting the expressions (2.32) into (2.31), we obtain

$$
\begin{aligned}
\tilde{y}' &= y' + \varepsilon\eta' + \cdots = \frac{d\tilde{y}}{d\tilde{x}} = \frac{dy + \varepsilon\, d\eta + \cdots}{dx + \varepsilon\, d\xi + \cdots} \\
&= \frac{y' + \varepsilon(d\eta/dx) + \cdots}{1 + \varepsilon(d\xi/dx) + \cdots} = y' + \varepsilon\left(\frac{d\eta}{dx} - y'\frac{d\xi}{dx}\right) + \cdots, \\
\tilde{y}^{(n)} &= y^{(n)} + \varepsilon\eta^{(n)} + \cdots = \frac{d\tilde{y}^{(n-1)}}{d\tilde{x}} \\
&= y^{(n)} + \varepsilon\left(\frac{d\eta^{(n-1)}}{dx} - y^{(n)}\frac{d\xi}{dx}\right) + \cdots,
\end{aligned}
\tag{2.34}
$$

from which we can read off the $\eta^{(i)}$:

$$
\eta' = \frac{d\eta}{dx} - y'\frac{d\xi}{dx} = \frac{\partial \eta}{\partial x} + y'\left(\frac{\partial \eta}{\partial y} - \frac{\partial \xi}{\partial x}\right) - y'^2\frac{\partial \xi}{\partial y}, \tag{2.35}
$$

$$
\eta^{(n)} = \frac{d\eta^{(n-1)}}{dx} - y^{(n)}\frac{d\xi}{dx}. \tag{2.36}
$$

It is easy to show by induction that the last recurrence formula (2.36) can also be written as

$$
\eta^{(n)} = \frac{d^n}{dx^n}(\eta - y'\xi) + y^{(n+1)}\xi. \tag{2.37}
$$

Note that $\eta^{(n)}$ is *not* the nth derivative of η!

We can summarize the result as follows: *if*

$$\mathbf{X}=\xi(x,y)\frac{\partial}{\partial x}+\eta(x,y)\frac{\partial}{\partial y} \tag{2.38}$$

is the infinitesimal generator of a point transformation, then

$$\mathbf{X}=\xi\frac{\partial}{\partial x}+\eta\frac{\partial}{\partial y}+\eta'\frac{\partial}{\partial y'}+\cdots+\eta^{(n)}\frac{\partial}{\partial y^{(n)}} \tag{2.39}$$

is its extension (prolongation) up to the nth *derivative, where the* $\eta^{(n)}(x,y,y',\ldots,y^{(n)})$ *are given by* (2.37).

In (2.38) and (2.39) we have used the same symbol **X** for the generator and its extension. This sloppiness in notation will not give rise to confusion, since in most cases it will be clear from the context what is meant. Only if we feel it necessary shall we denote the extension (2.39) by $\mathbf{X}^{(n)}$.

The extension formula (2.39) and the defining relation (2.37) for $\eta^{(n)}$ look rather simple, but this is a misleading impression. Since the derivatives in (2.37) are to be taken with respect to *all* arguments (chain rule!), the explicit expressions blow up enormously with increasing n, the first two steps being

$$\eta'=\eta_{,x}+(\eta_{,y}-\xi_{,x})y'-\xi_{,y}y'^2, \tag{2.40}$$

$$\begin{aligned}\eta''=&\,\eta_{,xx}+(2\eta_{,xy}-\xi_{,xx})y'+(\eta_{,yy}-2\xi_{,xy})y'^2\\&-\xi_{,yy}y'^3+(\eta_{,y}-2\xi_{,x}-3\xi_{,y}y')y''.\end{aligned} \tag{2.41}$$

Here – and in the following – we denote the partial derivative of a function by a comma followed by the variable with respect to which the derivation has been performed. We see from (2.37), (2.40), and (2.41) that the $\eta^{(n)}$ are *polynomials* in the derivatives $y',\ldots,y^{(n)}$ that are – for $n\geqslant 2$ – linear in the highest derivative $y^{(n)}$.

Some simple examples may illustrate the extension procedure. We always give the generator together with its extension up to the third derivative.

$$\begin{aligned}\mathbf{X}=&-y\frac{\partial}{\partial x}+x\frac{\partial}{\partial y}+(1+y'^2)\frac{\partial}{\partial y'}+3y'y''\frac{\partial}{\partial y''}\\&+(3y''^2+4y'y''')\frac{\partial}{\partial y'''}+\cdots,\end{aligned} \tag{2.42}$$

$$\mathbf{X}=x\frac{\partial}{\partial x}+y\frac{\partial}{\partial y}+0\frac{\partial}{\partial y'}-y''\frac{\partial}{\partial y''}-2y'''\frac{\partial}{\partial y'''}-\cdots, \tag{2.43}$$

$$\mathbf{X} = xy\frac{\partial}{\partial x} + y^2\frac{\partial}{\partial y} + (y - xy')y'\frac{\partial}{\partial y'} - 3xy'y''\frac{\partial}{\partial y''}$$

$$-(3y'y'' + 3xy''^2 + 4xy'y''' + yy''')\frac{\partial}{\partial y'''} + \cdots, \tag{2.44}$$

$$\mathbf{X} = \frac{\partial}{\partial x}. \tag{2.45}$$

This last example shows how to find the normal form of a generator extended to derivatives of arbitrary order: it suffices to transform the non-extended generator since with $\xi = 1$ and $\eta = 0$ (or vice versa) all $\eta^{(n)}$ vanish identically.

2.4 Multiple-parameter groups of transformations and their generators

Transformations can depend on more than one parameter ε; that is, instead of (2.2) we may have

$$\tilde{x} = \tilde{x}(x, y; \varepsilon_N), \qquad \tilde{y} = \tilde{y}(x, y; \varepsilon_N),$$
$$N = 1, \ldots, r. \tag{2.46}$$

If the ε_N are independent of each other, and if the transformations (2.46) contain the identity, are invertible, and include their repeated application (with possibly different ε_N), the transformations (2.46) form an r-parameter group G_r.

To each parameter ε_N, an infinitesimal generator $\mathbf{X}_N$ can be associated by

$$\mathbf{X}_N = \xi_N\frac{\partial}{\partial x} + \eta_N\frac{\partial}{\partial y},$$

$$\xi_N(x, y) = \left.\frac{\partial \tilde{x}}{\partial \varepsilon_N}\right|_{\varepsilon_M = 0}, \qquad \eta_N(x, y) = \left.\frac{\partial \tilde{y}}{\partial \varepsilon_N}\right|_{\varepsilon_M = 0}. \tag{2.47}$$

A rescaling of ε_N rescales the corresponding $\mathbf{X}_N$ by a constant factor [compare equation (2.11)], and a transformation

$$\varepsilon_N = \varepsilon_N(\hat{\varepsilon}_M), \qquad \varepsilon_N(0) = 0, \tag{2.48}$$

corresponds, because of, for example,

$$\hat{\xi}_N = \left.\frac{\partial \tilde{x}}{\partial \varepsilon_N}\right|_{\varepsilon_L = 0} = \left.\frac{\partial \tilde{x}}{\partial \varepsilon_M}\frac{\partial \varepsilon_M}{\partial \hat{\varepsilon}_N}\right|_{\varepsilon_L = 0} = \xi_M\left.\frac{\partial \varepsilon_M}{\partial \hat{\varepsilon}_N}\right|_{\varepsilon_L = 0} = B_N^M \xi_M, \tag{2.49}$$

to a linear transformation, with constant coefficients B_N^M, between the $\mathbf{X}_N$:

$$\hat{\mathbf{X}}_N = B_N^M \mathbf{X}_M. \tag{2.50}$$

If we want to specify a particular transformation, we have to say how the values of the different ε_A are related to each other. That is, we have to give the ε_A as functions of a single parameter ε. For this specific transformation the infinitesimal generator is

$$\xi = \left.\frac{\partial \tilde{x}}{\partial \varepsilon}\right|_{\varepsilon=0} = \left.\frac{\partial \tilde{x}}{\partial \varepsilon_N}\frac{\partial \varepsilon_N}{\partial \varepsilon}\right|_{\varepsilon=0} = a^N \xi_N, \quad \eta = a^N \eta_N,$$

$$\mathbf{X} = a^N \mathbf{X}_N; \tag{2.51}$$

that is, it is a linear combination (with constant coefficients a^N) of the basic generators $\mathbf{X}_N$.

If, for example, we have the two-parameter group of translations in the x–y plane

$$\tilde{x} = x + \varepsilon_1, \qquad \tilde{y} = y + \varepsilon_2, \tag{2.52}$$

then the generators are

$$\mathbf{X}_1 = \frac{\partial}{\partial x}, \qquad \mathbf{X}_2 = \frac{\partial}{\partial y}, \tag{2.53}$$

and a specific generator is given by

$$\mathbf{X} = a^1 \frac{\partial}{\partial x} + a^2 \frac{\partial}{\partial y}, \tag{2.54}$$

where the ratio of the constants a_1 and a_2 fixes a direction in the x–y plane (and for a specific translation we must also say how far to go in this direction).

Formally the only difference between the generator of a one-parameter group and that of a multiple-parameter group is that the latter contains some arbitrary parameters (the a^N) linearly. For such things as the normal form or extension of a single generator that does not matter at all, so that everything said in Sections 2.2 and 2.3 is valid here too. The more involved questions in connection with generators of multiple-parameter groups will be discussed in Chapter 6, where we shall deal with the properties of Lie algebras.

2.5 Exercises

1 Show that the formal solution

$$\tilde{x} = e^{\varepsilon \mathbf{X}} x \equiv x + \varepsilon \mathbf{X} x + \tfrac{1}{2}\varepsilon^2 \mathbf{X}(\mathbf{X}x) + \cdots, \qquad \tilde{y} = e^{\varepsilon \mathbf{X}} y,$$

of the system (2.10) coincides with the development with respect to powers of ε of the rotations (2.5) if applied to

$$\mathbf{X} = -y\frac{\partial}{\partial x} + x\frac{\partial}{\partial y}.$$

2 Show that to find the function $s(x, y)$ that one needs to transform a given generator $\mathbf{X}$ to its normal form $\mathbf{X} = \partial/\partial s$, one can proceed as follows. Solve

$$\frac{\mathrm{d}x}{\mathrm{d}s} = \xi[x(s), y(s)] \quad \text{and} \quad \frac{\mathrm{d}y}{\mathrm{d}s} = \eta[x(s), y(s)]$$

with initial values x_0, y_0, and invert these to obtain $s = s(x, y; x_0, y_0)$. Apply this procedure to

$$\mathbf{X} = x\frac{\partial}{\partial x} + y\frac{\partial}{\partial y}.$$

Why is the result not unique, and what degrees of freedom are there in the choice of $s(x, y)$?

3 The general projective transformation of the x–y plane is given by

$$\tilde{x} = \frac{ax + by + c}{ex + fy + g}, \qquad \tilde{y} = \frac{hx + ky + l}{ex + fy + g}.$$

Determine the generators $\mathbf{X}_N$. How many of them are linearly independent?

4 Do the transformations

$$\tilde{x} = x, \qquad \tilde{y} = ay + a^2 y^2, \qquad a = \text{const.}$$

form a group?

3

Lie point symmetries of ordinary differential equations: the basic definitions and properties

3.1 The definition of a symmetry: first formulation

A point transformation

$$\tilde{x} = \tilde{x}(x, y), \qquad \tilde{y} = \tilde{y}(x, y), \tag{3.1}$$

which may or may not depend on some parameters, is a symmetry transformation, or for short a symmetry, of an ordinary differential equation if it maps solutions into solutions; the image $\tilde{y}(\tilde{x})$ of any solution $y(x)$ is again a solution. That is to say, the nth order differential equation

$$H(x, y, y', \ldots, y^{(n)}) = 0 \tag{3.2}$$

does not change under a symmetry transformation; equations (3.1) and (3.2) imply

$$H(\tilde{x}, \tilde{y}, \tilde{y}', \ldots, \tilde{y}^{(n)}) = 0. \tag{3.3}$$

It is very important that this definition implies that the existence of a symmetry – of a mapping of solutions into solutions – is independent of the choice of variables in which the differential equation and its solutions are given. If we perform an arbitrary point transformation, only the explicit form of the symmetry transformation will change. We can expect simple differential equations to have many symmetries – so if we find several symmetries of a complicated looking differential equation, it may in fact be a simple one given in a disguised form, that is, in unsuitable variables.

Although symmetries that do *not* form a Lie group can be very useful in studying differential equations, there is no practicable way to find them. So from now on we shall assume that the symmetry transformation contains (at least) one parameter, ε,

$$\begin{aligned} &\tilde{x} = \tilde{x}(x, y; \varepsilon), \qquad \tilde{y} = \tilde{y}(x, y; \varepsilon), \\ &\tilde{y}' = \tilde{y}'(x, y, y'; \varepsilon), \quad \text{etc.,} \end{aligned} \tag{3.4}$$

compare Section 2.3. We then call it a Lie point symmetry.

We now want to give the symmetry condition an analytic form. Since (3.3) has to be valid for all values of ε, we obtain from (3.3) by differentiation

$$\begin{aligned} 0 &= \left.\frac{\partial H(\tilde{x}, \tilde{y}, \ldots, \tilde{y}^{(n)})}{\partial \varepsilon}\right|_{\varepsilon=0} \\ &= \left.\left(\frac{\partial H}{\partial \tilde{x}}\frac{\partial \tilde{x}}{\partial \varepsilon} + \frac{\partial H}{\partial \tilde{y}}\frac{\partial \tilde{y}}{\partial \varepsilon} + \cdots + \frac{\partial H}{\partial \tilde{y}^{(n)}}\frac{\partial \tilde{y}^{(n)}}{\partial \varepsilon}\right)\right|_{\varepsilon=0}. \end{aligned} \tag{3.5}$$

Because, for example, $(\partial H/\partial \tilde{x})|_{\varepsilon=0} = (\partial H/\partial x)$ and using the definitions (2.8) and (2.33) of $\xi, \eta, \eta', \ldots$, this is equivalent to

$$\xi\frac{\partial H}{\partial x} + \eta\frac{\partial H}{\partial y} + \eta'\frac{\partial H}{\partial y'} + \cdots + \eta^{(n)}\frac{\partial H}{\partial y^{(n)}} = 0 \tag{3.6}$$

or to

$$\mathbf{X}H = 0. \tag{3.7}$$

If the differential equation $H = 0$ holds and admits a group of symmetries with generator(s) $\mathbf{X}$, then $\mathbf{X}H = 0$ also holds; $H = 0$ is invariant under the infinitesimal transformation.

To prove the converse, we shall make use of the important fact that the existence of a symmetry is independent of the choice of variables; only its explicit form (the components of $\mathbf{X}$) depend on this choice; compare Section 2.2. So if $\mathbf{X}H = 0$ holds for some $\mathbf{X}$ (and $H = 0$), we can always transform this generator to its normal form $\mathbf{X} = \partial/\partial s$. Choosing s as the new independent variable $\tilde{x}$ and dropping the tilde, equation (3.7) then reads

$$\mathbf{X}H = \frac{\partial H}{\partial x} = 0. \tag{3.8}$$

If this relation were true in general, we could conclude that H does not depend on x. But since (3.8) need be satisfied only for solutions of $H = 0$, this deduction is not always possible. To see the reason, take, for example, the differential equation

$$H = (y'' - x)^2 = 0. \tag{3.9}$$

For this differential equation, all first derivatives of H vanish on $H = 0$; $\mathbf{X}H = 0$ holds for any linear operator $\mathbf{X}$. In particular, that $\partial H/\partial x = 0$ for solutions of $H = 0$ does *not* imply that H is independent of x. To get rid of this singular behaviour of the differential equation, we have to impose the restriction that not all first derivatives of H vanish on $H = 0$. This can be done by assuming that the differential equation can be solved for the highest derivative and can thus be written as

$$H = y^{(n)} - \omega(x, y, y', \ldots, y^{(n-1)}) = 0. \tag{3.10}$$

Then $\mathbf{X}H = \partial H/\partial x = \partial\omega/\partial x = 0$ on $H = 0$ indeed implies that the function H does not depend on x. So the finite transformation $\tilde{x} = x + \varepsilon$ does not change H; it is a symmetry of the differential equation.

To summarize, we have arrived at the following theorem (which is valid if not all first derivatives of H vanish on $H = 0$).

An ordinary differential equation

$$H(x, y, y', \ldots, y^{(n)}) = 0 \tag{3.11}$$

admits a group of symmetries with generator $\mathbf{X}$ *if and only if*

$$\mathbf{X}H \equiv 0 \qquad (\text{mod } H = 0) \tag{3.12}$$

holds.

This last equation (3.12) needs some explanation: why "$\equiv 0$", and why "mod $H = 0$"? We have chosen to write $\mathbf{X}H \equiv 0$ instead of $\mathbf{X}H = 0$ to stress the fact that this is an equation that has to be satisfied *identically* in all the variables $x, y, y', \ldots, y^{(n-1)}$. This reflects the property of a symmetry that (3.12) has to be true for *every* solution $y(x)$ of (3.11), and for an arbitrary solution the (initial) values $y, y', \ldots, y^{(n-1)}$ can be prescribed freely at any point x. "Mod $H = 0$" indicates, of course, that the highest derivative $y^{(n)}$ is fixed by the differential equation $H = 0$ if $x, y, \ldots, y^{(n-1)}$ are given: $y^{(n)}$ should be eliminated – by means of $H = 0$ – from $\mathbf{X}H = 0$ before this equation is evaluated.

For example, in the linear differential equation

$$H(x, y, y', y'') = y'' + y = 0 \tag{3.13}$$

y (and y'') can be multiplied by an arbitrary constant factor; that is, (3.13) admits the symmetry

$$\tilde{x} = x, \qquad \tilde{y} = (1+\varepsilon)y, \qquad \mathbf{X} = y\frac{\partial}{\partial y} + y'\frac{\partial}{\partial y'} + y''\frac{\partial}{\partial y''}. \tag{3.14}$$

This yields

$$\mathbf{X}H = y + y'', \tag{3.15}$$

and $\mathbf{X}H$ vanishes only if $H = 0$ is taken into account.

For a number of applications a different (although equivalent) formulation and definition of a symmetry will prove useful. For this formulation, we need some knowledge about the connections between ordinary differential equations and linear partial differential equations of first order, which is provided in the next section.

3.2 Ordinary differential equations and linear partial differential equations of first order

There is a close connection between ordinary differential equations of nth order and first order linear partial differential equations in $n+1$ variables, which we shall now establish. To do this, we first need some facts about such partial differential equations.

The partial differential equations in question can be written as

$$\mathbf{A}f = a^i\frac{\partial}{\partial x^i}f = a^i(x^r)\frac{\partial}{\partial x^i}f(x^r) = 0, \qquad i, r = 1, \ldots, n+1. \tag{3.16}$$

Obviously $\mathbf{A}f = 0$ and $\lambda(x^k)\mathbf{A}f = 0$ will have the same set of solutions, and if f is a solution, then an arbitrary function $F(f)$ is a solution too.

As already stated in the context of the operator $\mathbf{X}$ in Section 2.2, a linear operator

$$\mathbf{A} = a^i\frac{\partial}{\partial x^i} \tag{3.17}$$

can always be transformed to its normal form

$$\mathbf{A} = \frac{\partial}{\partial s} \tag{3.18}$$

by introducing appropriate coordinates

$$s = s(x^i), \qquad \varphi^\alpha = \varphi^\alpha(x^i), \qquad \alpha = 1, \ldots, n, \tag{3.19}$$

with

$$\mathbf{A}\varphi^\alpha = 0. \tag{3.20}$$

Neither s nor the φ^α is uniquely determined; transformations $s = s + S(\varphi^\alpha)$ and $\hat{\varphi}^\alpha = \hat{\varphi}^\alpha(\varphi^\beta)$ are always possible without violating (3.18) and (3.20). But we can infer from (3.17)–(3.20) that the number of functionally independent solutions of (3.16) equals the number of functionally independent functions φ^α, which is exactly n. By the way, solving $\mathbf{A}f = 0$ and transforming $\mathbf{A}$ to its normal form (3.18) are the same problem!

Conversely, if in the space of $n+1$ coordinates a system of n functionally independent functions $\varphi^\alpha(x^i)$ that are solutions to an equation $\mathbf{A}f = 0$ is given, then $\mathbf{A}$ is determined by the φ^α up to a (nonconstant) factor λ: if one uses φ^α and some function $s(x^i)$ as coordinates, then $\mathbf{A}$ must be proportional to $\partial/\partial s$.

We are now ready to establish the correspondence between ordinary differential equations and partial differential equations of the type (3.16). To do this, we write the ordinary differential equation of nth order as

$$y^{(n)} = \omega(x, y, y', \ldots, y^{(n-1)}) \tag{3.21}$$

and assign to it the partial differential equation in $n+1$ variables,

$$\mathbf{A}f = \left(\frac{\partial}{\partial x} + y'\frac{\partial}{\partial y} + y''\frac{\partial}{\partial y'} + \cdots + \omega\frac{\partial}{\partial y^{(n-1)}}\right)f = 0. \tag{3.22}$$

Note that in (3.22) the quantities $y', y'', \ldots, y^{(n-1)}$ are treated as independent variables on the same footing as x and y!

The formal relation between the two equations (3.21) and (3.22) rests on the function ω occurring in both. The deeper relation between them is that they are completely equivalent, the link being provided by the *first integrals* of the ordinary differential equation. A first integral is a function $z = z(x, y, y', \ldots, y^{(n-1)})$ that is constant along the solutions of (3.21); that is,

$$\frac{\mathrm{d}z}{\mathrm{d}x} = \frac{\partial z}{\partial x} + y'\frac{\partial z}{\partial y} + y''\frac{\partial z}{\partial y'} + \cdots + y^{(n)}\frac{\partial z}{\partial y^{(n-1)}} = 0 \tag{3.23}$$

holds if $y^{(n)}$ is substituted by ω. Since, by inverting the relation $z(x, y, y', \ldots, y^{(n-1)}) = \text{const.} = z_0$ with respect to $y^{(n-1)}$, we obtain

$$y^{(n-1)} = \hat{\omega}(x, y, \ldots, y^{(n-2)}; z_0), \tag{3.24}$$

the first integral can be used to reduce the order of the differential equation by one. Comparing the definition (3.23) of a first integral with the partial differential equation (3.22), one sees immediately that every solution φ^α of

$\mathbf{A}f = 0$ is a first integral z of $y^{(n)} = \omega$, and conversely (note that φ^α must depend on $y^{(n-1)}$ since $\partial\varphi^\alpha/\partial y^{(n-1)} = 0$ contradicts $\mathbf{A}\varphi^\alpha = 0$ unless φ^α is constant). Moreover, every complete set of n functionally independent solutions φ^α corresponds to the general solution $y = y(x, \varphi_0^\alpha)$ of the ordinary differential equation that can be obtained by eliminating all derivatives of y from the system

$$\varphi^\alpha(x, y, y', \ldots, y^{(n-1)}) = \varphi_0^\alpha, \qquad \alpha = 1, \ldots, n. \tag{3.25}$$

The constants φ_0^α are essentially the constants of integration (initial values) of the differential equation.

To illustrate this theoretical framework, let us consider the simple differential equation

$$y'' = \omega(x, y, y') = -y \tag{3.26}$$

and its associated partial differential equation

$$\mathbf{A}f = \left(\frac{\partial}{\partial x} + y'\frac{\partial}{\partial y} - y\frac{\partial}{\partial y'}\right)f = 0. \tag{3.27}$$

Two functionally independent solutions φ^1 and φ^2 of $\mathbf{A}f = 0$ are provided by

$$\varphi^1 = y^2 + y'^2, \qquad \varphi^2 = x - \arctan y/y', \tag{3.28}$$

and from $\varphi^1 = \varphi_0^1 = \text{const.}$, $\varphi^2 = \varphi_0^2 = \text{const.}$, one can obtain the general solution of (3.26) as

$$y = y(x; \varphi_0^1, \varphi_0^2) = (\varphi_0^1)^{1/2}\sin(x - \varphi_0^2). \tag{3.29}$$

We must now show that the substitution of an ordinary by a partial differential equation makes sense in the context of symmetries. This will be done immediately.

3.3 The definition of a symmetry: second formulation

Since, as shown above, the differential equation $y^{(n)} = \omega$ can be expressed by means of the linear operator $\mathbf{A}$ as

$$\mathbf{A}f = \left(\frac{\partial}{\partial x} + y'\frac{\partial}{\partial y} + y''\frac{\partial}{\partial y'} + \cdots + \omega\frac{\partial}{\partial y^{(n-1)}}\right)f = 0, \tag{3.30}$$

and any Lie point symmetry that may exist is fully determined by its generator

$$\mathbf{X} = \xi\frac{\partial}{\partial x} + \eta\frac{\partial}{\partial y} + \eta'\frac{\partial}{\partial y'} + \cdots + \eta^{(n-1)}\frac{\partial}{\partial y^{(n-1)}}, \tag{3.31}$$

the differential equation (operator **A**) and symmetry (generator **X**) are on an equal footing!

What are the conditions the generator **X** has to satisfy to be a symmetry of (3.30)? Take a (complete) set of n independent solutions φ^α of (3.30). By definition, a symmetry – and hence its generator – maps solutions into solutions; $\mathbf{X}\varphi^\alpha$ must be again a solution, and we have

$$\mathbf{X}\varphi^\alpha = \Omega^\alpha(\varphi^\beta), \qquad \mathbf{A}\varphi^\alpha = 0 = \mathbf{A}\Omega^\alpha. \tag{3.32}$$

To get rid of the unknown functions Ω^α, we consider the commutator of **X** and **A**, which is defined by

$$[\mathbf{X}, \mathbf{A}] = \mathbf{XA} - \mathbf{AX} = -(\mathbf{A}\xi)\frac{\partial}{\partial x} + [(\mathbf{X}y') - (\mathbf{A}\eta)]\frac{\partial}{\partial y} + \cdots$$
$$+ [(\mathbf{X}\omega) - (\mathbf{A}\eta^{(n-1)})]\frac{\partial}{\partial y^{(n-1)}} \tag{3.33}$$

and is again a linear operator like **A** and **X**.

Because of (3.32), we obtain

$$[\mathbf{X}, \mathbf{A}]\varphi^\alpha = \mathbf{X}(\mathbf{A}\varphi^\alpha) - \mathbf{A}(\mathbf{X}\varphi^\alpha) = 0. \tag{3.34}$$

Since this is valid for all functions φ^α, the equation $[\mathbf{X}, \mathbf{A}]f = 0$ has the same set of solutions as the equation $\mathbf{A}f = 0$, so that these two operators can differ only by a (in general, nonconstant) factor λ:

$$[\mathbf{X}, \mathbf{A}] = \lambda(x, y, \ldots, y^{(n-1)})\mathbf{A}. \tag{3.35}$$

If the condition (3.35) is satisfied, then **X** is a symmetry of $\mathbf{A}f = 0$. Note that $[\mathbf{X}, \mathbf{A}]$ and **A** are essentially vectors, so that (3.35) is a system of $n+1$ equations. At first glance this number $n+1$ may be surprising since the competing definition [(3.12)] $\mathbf{X}H = 0$ is only *one* equation. To clarify this point, let us consider the different components of (3.35); that is, let us compare the coefficients of $\partial/\partial x$, $\partial/\partial y$, $\partial/\partial y'$,... on the two sides of this equation, making use, of course, of relation (3.33).

Comparing the coefficients of $\partial/\partial x$ yields

$$-\mathbf{A}\xi = -\left(\frac{\partial \xi}{\partial x} + y'\frac{\partial \xi}{\partial y}\right) = \lambda, \tag{3.36}$$

which is obviously the definition of the function λ. Before going on to the rest of the equations (3.35), we observe that although the operator **A** as defined by (3.22) treats the derivatives of y as independent variables, it can formally be written as

$$\mathbf{A} = \frac{\mathrm{d}}{\mathrm{d}x} \qquad (\mathrm{mod}\ y^{(n)} = \omega), \tag{3.37}$$

where the mod $y^{(n)} = \omega$ indicates that $y^{(n)}$ has to be substituted by $\omega(x, y, \ldots, y^{(n-1)})$ whenever it occurs. Because of (3.31), (3.33), and (3.36), equation (3.35) then reads

$$\left(\eta' - \frac{\mathrm{d}\eta}{\mathrm{d}x}\right)\frac{\partial}{\partial y} + \left(\eta'' - \frac{\mathrm{d}\eta'}{\mathrm{d}x}\right)\frac{\partial}{\partial y'} + \cdots + \left(\mathbf{X}\omega - \frac{\mathrm{d}\eta^{(n-1)}}{\mathrm{d}x}\right)\frac{\partial}{\partial y^{(n-1)}}$$
$$= -\frac{\mathrm{d}\xi}{\mathrm{d}x}\left(y'\frac{\partial}{\partial y} + y''\frac{\partial}{\partial y'} + \cdots + \omega\frac{\partial}{\partial y^{(n-1)}}\right) \qquad (\mathrm{mod}\ y^{(n)} = \omega), \tag{3.38}$$

where the mod $y^{(n)} = \omega$ in practice concerns only the coefficient of $\partial/\partial y^{(n-1)}$ on the left side. We now see that because of the definition (2.36) of the $\eta^{(i)}$, most terms cancel, with the exception of the coefficients of $\partial/\partial y^{(n-1)}$, which lead to

$$\mathbf{X}\omega = \eta^{(n)} \qquad (\mathrm{mod}\ y^{(n)} = \omega). \tag{3.39}$$

But this is exactly the result we would have obtained from the alternative symmetry condition (3.12), that is, from $\mathbf{X}H = 0$ (mod $H = 0$) by inserting $H = y^{(n)} - \omega$. So the two approaches to the problem and the two definitions (3.12) and (3.35) of a symmetry are completely equivalent. The advantage of formulating the symmetry condition with the help of the linear operator **A** will become clear in the later chapters dealing with the application of symmetries.

One important remark should be added to the calculations and reasoning of the last paragraph. We could have started with the form (3.31) of the generator **X** *without* specifying the functions $\eta^{(i)}$. The symmetry condition $[\mathbf{X}, \mathbf{A}] = \lambda\mathbf{A}$ [in the form, e.g., of equation (3.38)] would then give us the correct expressions for these functions! So the symmetry condition (3.35) includes the rule for extending point transformations to the derivatives; in this respect it is more general than the condition $\mathbf{X}H \equiv 0$ (mod $H = 0$). When generalizing the class of transformations and generators, we shall therefore always start from the form $[\mathbf{X}, \mathbf{A}] = \lambda\mathbf{A}$ of the symmetry condition.

3.4 Summary

The nth order differential equation can be given in any of the forms

$$H(x, y, y', \ldots, y^{(n)}) = 0 \tag{3.40}$$

or

$$y^{(n)} = \omega(x, y, y', \ldots, y^{(n-1)}) \tag{3.41}$$

or

$$\mathbf{A}f = \left(\frac{\partial}{\partial x} + y'\frac{\partial}{\partial y} + y''\frac{\partial}{\partial y'} + \cdots + \omega\frac{\partial}{\partial y^{(n-1)}}\right)f = 0. \tag{3.42}$$

This differential equation admits a Lie point symmetry with generator

$$\mathbf{X} = \xi(x, y)\frac{\partial}{\partial x} + \eta(x, y)\frac{\partial}{\partial y} + \eta'\frac{\partial}{\partial y'} + \cdots + \eta^{(i)}\frac{\partial}{\partial y^{(i)}},$$

$$\eta^{(m)} = \frac{\mathrm{d}\eta^{(m-1)}}{\mathrm{d}x} - y^{(m)}\frac{\mathrm{d}\xi}{\mathrm{d}x}, \tag{3.43}$$

if

$$\mathbf{X}H = 0 \qquad (\mathrm{mod}\, H = 0) \tag{3.44}$$

or (equivalently)

$$[\mathbf{X}, \mathbf{A}] = \lambda\mathbf{A} \tag{3.45}$$

holds.

3.5 Exercises

1 Suppose you know the general solution $y = y(x, a^\alpha)$, $\alpha = 1, \ldots, n$, of an nth order differential equation. How can you obtain from it n independent solutions φ^α of $\mathbf{A}f = 0$?

2 Let $\mathbf{X}$ be a Lie point symmetry of $\mathbf{A}$, that is, $[\mathbf{X}, \mathbf{A}] = \lambda\mathbf{A}$. Define $\hat{\mathbf{X}}$ by $\hat{\mathbf{X}} = \mathbf{X} + \mu(x, y, y', \ldots, y^{(n-1)})\mathbf{A}$. Then $[\hat{\mathbf{X}}, \mathbf{A}] = \hat{\lambda}\mathbf{A}$ holds with some $\hat{\lambda}$. Is $\hat{\mathbf{X}}$ again a Lie point symmetry of $\mathbf{A}$?

3 Two operators $\mathbf{A}$ and $\hat{\mathbf{A}} = \mu(x, y, y', \ldots, y^{(n-1)})\mathbf{A}$ are equivalent since they have the same set of solutions φ^α. Show that if $\mathbf{X}$ is a symmetry of $\mathbf{A}$, then there is always an equivalent $\hat{\mathbf{A}}$ that commutes with $\mathbf{X}$.

4 Show by using the definition $[\mathbf{X}, \mathbf{A}] = \lambda\mathbf{A}$ of a symmetry that the operations "extension of a generator" and "point transformation" commute, that is, that $\widetilde{\eta^{(n)}} = \tilde{\eta}^{(n)}$ holds.

4

How to find the Lie point symmetries of an ordinary differential equation

4.1 Remarks on the general procedure

Finding the Lie point symmetries of a differential equation $H = 0$ means finding the general solution $\xi(x, y), \eta(x, y)$ of the symmetry condition (3.44). To ensure the regularity of the differential equation $H = 0$ discussed in Section 3.1, and since many differential equations naturally arise as equations linear in the highest derivative, we prefer to start from the form $y^{(n)} = \omega(x, y, y', \ldots, y^{(n-1)})$ of the differential equation. Because of (3.39), the symmetry condition reads

$$\mathbf{X}\omega = \left(\xi\frac{\partial}{\partial x} + \eta\frac{\partial}{\partial y} + \eta'\frac{\partial}{\partial y'} + \cdots + \eta^{(n-1)}\frac{\partial}{\partial y^{(n-1)}}\right)\omega = \eta^{(n)}, \tag{4.1}$$

with $\eta^{(i)}$ given by

$$\eta^{(i)} = \frac{\mathrm{d}\eta^{(i-1)}}{\mathrm{d}x} - y^{(i)}\frac{\mathrm{d}\xi}{\mathrm{d}x}, \tag{4.2}$$

where the $y^{(n)}$ appearing (only) in $\eta^{(n)}$ must be substituted by ω.

The condition (4.1) is in fact a differential equation for the functions ξ and η that is *linear* in both functions. Moreover, since the condition has to be satisfied *identically* in all variables $x, y, y', \ldots, y^{(n-1)}$ but ξ and η are functions of x and y only, it will split into many partial differential equations whose solution (possibly trivial) can usually be found.

So start by writing down the condition (4.1) in detail. Then collect, for example, all terms containing the highest derivative $y^{(n-1)}$ and make them vanish identically; this usually gives several partial differential equations. Try to solve these equations and then turn to the terms with, for example, $y^{(n-2)}$. Even if at first glance the task seems to be very complicated, simply go ahead; where symmetries exist, they can usually be determined, but to be honest, in many cases it will turn out there are none.

There are two important exceptions to this optimistic view of the problem. These are the first order differential equations and the general linear differential equations. We shall come to these cases immediately.

4.2 The atypical case: first order differential equations

For a first order differential equation

$$y' = \omega(x, y), \qquad \mathbf{A} = \frac{\partial}{\partial x} + \omega\frac{\partial}{\partial y}, \tag{4.3}$$

the symmetry condition reads

$$\mathbf{X}\omega = \left(\xi\frac{\partial}{\partial x} + \eta\frac{\partial}{\partial y}\right)\omega = \omega\left(\frac{\partial\eta}{\partial y} - \omega\frac{\partial\xi}{\partial y} - \frac{\partial\xi}{\partial x}\right) + \frac{\partial\eta}{\partial x} \tag{4.4}$$

or

$$\xi\omega_{,x} + \xi_{,x}\omega + \xi_{,y}\omega^2 = \eta_{,x} + \eta_{,y}\omega - \eta\omega_{,y}. \tag{4.5}$$

The function $\omega(x, y)$ being given, this partial differential equation always has (nonzero) solutions $\xi(x, y)$ and $\eta(x, y)$. One can even prescribe η (or ξ) and then determine ξ (or η) from (4.5). A first order differential equation always has an infinite number of symmetries!

One can show this also by a simple geometric argument. The set of solutions to the differential equation is a one-parameter family of curves (see Figure 4.1). A symmetry operation is by definition a transformation that maps solutions into solutions, and its generator **X** corresponds to a vector field leading from solutions (curves) to their (infinitesimally) neighbouring solutions (curves). But such a vector field always exists! One can even prescribe the direction of **X** at any point (note that **X** is defined only up to a constant factor). Two different choices $\hat{\mathbf{X}}$ and **X** differ by a motion in the direction of the integral curves. Analytically, if **X** is a symmetry, then $\hat{\mathbf{X}} = \mathbf{X} + \mu(x, y)\mathbf{A}$ is too, since $[\hat{\mathbf{X}}, \mathbf{A}] = \hat{\lambda}\mathbf{A}$ holds and $\hat{\xi}, \hat{\eta}$ are functions only of x and y (compare Exercise 3.2).

Although an infinite number of solutions to (4.5) exist, there is no systematic way to find even one of them; this property characterizes the atypical behaviour of first order differential equations.

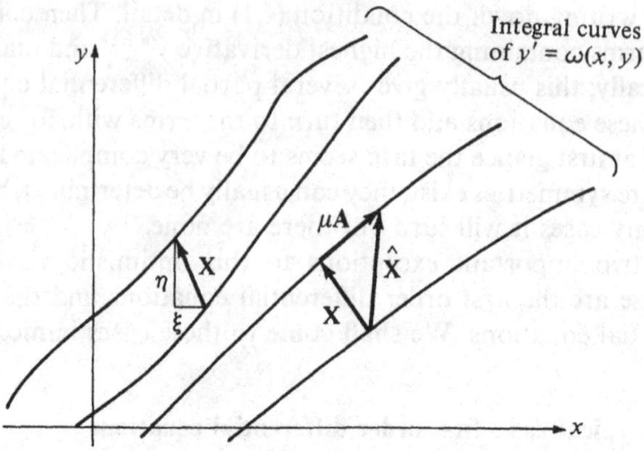

Figure 4.1. The symmetries of a first order differential equation.

Nevertheless, simple cases exist where a solution to the symmetry condition (4.5) can be given. We list a few of them, together with their symmetry generators:

$$y' = -(y + 2x)/x, \qquad \mathbf{X} = \frac{1}{y + 2x}\frac{\partial}{\partial x}, \tag{4.6}$$

$$y' = f(y)g(x), \qquad \mathbf{X} = \frac{1}{g(x)}\frac{\partial}{\partial x}, \tag{4.7}$$

$$y' = 1 + \frac{1}{x}\tan(x - y), \qquad \mathbf{X} = \frac{1}{x\cos(x - y)}\frac{\partial}{\partial y}. \tag{4.8}$$

4.3 Second order differential equations

For a second order differential equation

$$y'' = \omega(x, y, y') \tag{4.9}$$

the symmetry condition (4.1) reads

$$\mathbf{X}\omega = \left(\xi\frac{\partial}{\partial x} + \eta\frac{\partial}{\partial y} + \eta'\frac{\partial}{\partial y'}\right)\omega = \eta''. \tag{4.10}$$

If we use the expressions (2.40) and (2.41) for η' and η'' (and, of course, substitute y'' by ω), we obtain

$$\begin{aligned}\omega(\eta_{,y} - 2\xi_{,x} - 3y'\xi_{,y}) - \omega_{,x}\xi - \omega_{,y}\eta - \omega_{,y'}[\eta_{,x} + y'(\eta_{,y} - \xi_{,x}) - y'^2\xi_{,y}]\\ + \eta_{,xx} + y'(2\eta_{,xy} - \xi_{,xx}) + y'^2(\eta_{,yy} - 2\xi_{,xy}) - y'^3\xi_{,yy} = 0.\end{aligned} \tag{4.11}$$

From this differential equation we have to determine $\xi(x, y)$ and $\eta(x, y)$. As already stressed in Section 4.1, it is an identity in x, y, *and* y', and since ξ and η do not depend on y', (4.11) will split into several equations according to the different dependence of its parts on y'.

We now illustrate the general procedure by the example of the differential equation

$$y'' = \omega(x, y, y') = x^n y^2, \qquad n \neq 0. \tag{4.12}$$

The symmetry condition (4.11) here reads

$$\begin{aligned} &x^n y^2(\eta_{,y} - 2\xi_{,x}) - nx^{n-1}y^2\xi - 2yx^n\eta + \eta_{,xx} + y'(2\eta_{,xy} - \xi_{,xx} - 3x^n y^2 \xi_{,y}) \\ &\quad + y'^2(\eta_{,yy} - 2\xi_{,xy}) - y'^3\xi_{,yy} = 0. \end{aligned} \tag{4.13}$$

Equating to zero the coefficients of y'^3 and y'^2 yields

$$\xi_{,yy} = 0, \qquad \eta_{,yy} = 2\xi_{,xy}, \tag{4.14}$$

which integrates to give

$$\xi = y\alpha(x) + \beta(x), \qquad \eta = y^2\alpha'(x) + y\gamma(x) + \delta(x). \tag{4.15}$$

Note that ξ and η are polynomials in y, so that the parts of (4.13) still to be satisfied will split again into several equations corresponding to the different powers of y that occur! From equating to zero the coefficient of y', we obtain

$$-3x^n y^2\alpha + 3y\alpha'' + 2\gamma' - \beta'' = 0, \tag{4.16}$$

which leads to $\alpha = 0$ and $2\gamma' = \beta''$. So we have

$$\xi = \beta(x), \qquad \eta = y(\tfrac{1}{2}\beta' + c) + \delta(x), \qquad c = \text{const.} \tag{4.17}$$

On inserting this into (4.13) – only terms without y' need to be considered – we finally obtain

$$-y^2[x^n(\tfrac{5}{2}\beta' + c) + nx^{n-1}\beta] + y(\tfrac{1}{2}\beta''' - 2x^n\delta) + \delta'' = 0, \tag{4.18}$$

that is,

$$\tfrac{5}{2}x\beta' + n\beta + cx = 0, \qquad \delta = \tfrac{1}{4}x^{-n}\beta''', \qquad \delta'' = 0. \tag{4.19}$$

From the first of these equations we can determine β, the second gives us δ, and the last equation is then a condition to be satisfied by the solution β of

Table 4.1. *Symmetries of the differential equation* $y'' = x^n y^2$ $(n \neq 0)$

$n = -5$:	$\mathbf{X} = (ax^2 + bx)\frac{\partial}{\partial x} + y(ax + 3b)\frac{\partial}{\partial y}$
$n = -\frac{15}{7}$:	$\mathbf{X} = \left(a\frac{7^3}{12}x^{6/7} + 7bx\right)\frac{\partial}{\partial x} + [a + y(b + \frac{49}{4}ax^{-1/7})]\frac{\partial}{\partial y}$
$n = -\frac{20}{7}$:	$\mathbf{X} = \left(a\frac{7^3}{12}x^{8/7} + \frac{7}{6}bx\right)\frac{\partial}{\partial x} + [-ax + y(b + \frac{49}{3}ax^{1/7})]\frac{\partial}{\partial y}$
all other n:	$\mathbf{X} = bx\frac{\partial}{\partial x} - (n + 2)by\frac{\partial}{\partial y}$

the first one. When going through this program, it will turn out that several cases have to be distinguished according to the value of n. The results are listed in Table 4.1.

We see that the differential equation $y'' = x^n y^2$ always admits one symmetry, and in some exceptional cases two symmetries (two parameters a and b).

If we try to apply the above method of solving the symmetry condition to the second differential equation mentioned in the introduction, namely to

$$y'' = (x - y)y'^3, \tag{4.20}$$

we encounter more difficulties. The symmetry condition (4.11) now reads

$$\begin{aligned}&(x - y)y'^3(\eta_{,y} - 2\xi_{,x} - 3\xi_{,y}y') - \xi y'^3 + \eta y'^3\\&- 3y'^2(x - y)[\eta_{,x} + y'(\eta_{,y} - \xi_{,x}) - y'^2\xi_{,y}] + \eta_{,xx}\\&+ y'(2\eta_{,xy} - \xi_{,xx}) + y'^2(\eta_{,yy} - 2\xi_{,xy}) - y'^3\xi_{,yy} = 0.\end{aligned} \tag{4.21}$$

From equating to zero the terms without y' and those linear in y', we obtain $\eta_{,xx} = 0$ and $2\eta_{,xy} - \xi_{,xx} = 0$, that is,

$$\xi = \alpha'(y)x^2 + \gamma(y)x + \delta(y), \qquad \eta = \alpha(y)x + \beta(y). \tag{4.22}$$

The remaining terms of the condition then yield

$$\begin{aligned}&\alpha'' + \alpha = 0,\\&2\gamma' - \beta'' - 3y\alpha = 0,\end{aligned} \tag{4.23}$$

$$2\beta' + \gamma'' - \alpha = 0,$$

$$\delta'' + \delta + y(\gamma - 2\beta') - \beta = 0.$$

Although these equations can be solved explicitly, it is a little bit awkward to do so. The strategy, of course, is to begin with the first equation to get $\alpha(y)$ (two constants of integration); insert α into the second equation and integrate once to get $\gamma(y)$ in terms of β' (one constant); then determine β from the third equation, which is of third order (three constants); and finally solve the last equation for δ (two constants). Altogether eight constants of integration appear; the differential equation (4.20) admits an eight-parameter group of point symmetries!

Two particular simple solutions of (4.23) that can be found by inspection are $\beta = a = \text{const.} = \delta$ $(\gamma = \alpha = 0)$ and $\gamma = \text{const.} = b = -\delta'$ $(\alpha = \beta = 0)$; the symmetries of $y'' = (x - y)y'^3$ contain among others

$$\mathbf{X} = [a + b(x - y)]\frac{\partial}{\partial x} + a\frac{\partial}{\partial y}. \tag{4.24}$$

It will become clear later (in Section 4.4) that the relative difficulties in finding all symmetries and the high number of symmetries are not unrelated to each other.

The simplest second order differential equation is, of course,

$$y'' = 0, \tag{4.25}$$

and we expect it to have many (and simple) symmetries. Indeed, for $\omega = 0$ the symmetry condition (4.11) immediately leads to $\eta_{,xx} = 2\eta_{,xy} - \xi_{,xx} = \eta_{,yy} - 2\xi_{,xy} = \xi_{,yy} = 0$, which are easy to solve and give

$$\mathbf{X} = (a_1 + a_2 x + a_3 y + a_4 xy + a_5 x^2)\frac{\partial}{\partial x}$$

$$+ (a_6 + a_7 x + a_8 y + a_5 xy + a_4 y^2)\frac{\partial}{\partial y}. \tag{4.26}$$

This is exactly the eight-parameter generator of the general projective transformation of Exercise 2.3.

Can a second order differential equation admit more than eight symmetries? What is the maximum number of symmetries? To answer these questions, assume that the equation admits a group with (at least) nine parameters; that is, the generator can be written as

$$\mathbf{X} = a^N \mathbf{X}_N, \qquad N = 1, \ldots, 9. \tag{4.27}$$

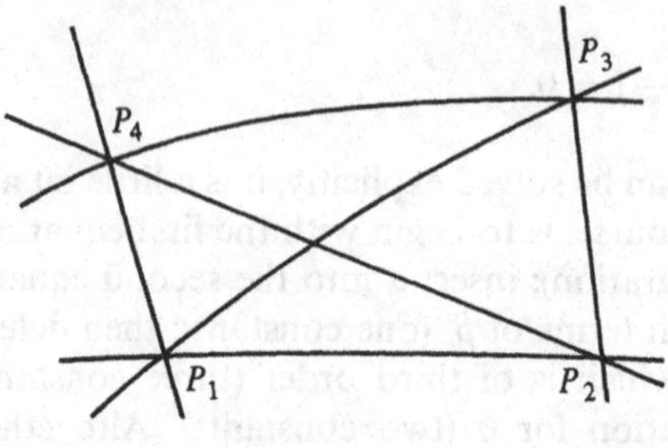

Figure 4.2. Fixed points and invariant solutions of an assumed nine-parameter group of symmetries.

A second order differential equation has the property that – at least locally – a solution is uniquely fixed by prescribing y' at some point (x, y) or prescribing two points P_1, P_2 through which to draw a solution curve. Consider now four points $P_1, \ldots, P_4$ in a general position. They uniquely define a set of six curves, three through each point (see Figure 4.2). We can then choose the nine parameters a^N in (4.27) so that all four points $P_1, \ldots, P_4$ remain at rest; that is, so that

$$\xi(P_i) = a^N \xi_N(P_i) = 0, \qquad \eta(P_i) = a^N \eta_N(P_i) = 0,$$
$$i = 1, \ldots, 4, \tag{4.28}$$

holds (this is a system of eight homogeneous linear equations for the a^N, and the $\mathbf{X}_N$ were assumed to be linearly independent). But when the P_i remain at rest, all integral curves connecting them must be mapped into themselves, and that implies that at each of the four points at least three directions y' (those of the three integral curves in question) remain unchanged. That means that at each of the P_i the value of $\eta'(P_i)$ is zero for these three values of y'.

Because of the definition (2.35), we have

$$\eta'(P_i) = \frac{\partial \eta}{\partial x} + y'\left(\frac{\partial \eta}{\partial y} - \frac{\partial \xi}{\partial x}\right) + y'^2 \frac{\partial \xi}{\partial y}\bigg|_{P_i}$$
$$= c_i + d_i y' + e_i y'^2 = 0, \tag{4.29}$$

and if *three* different solutions y' to these equations exist for each P_i, all the coefficients c_i, d_i, e_i must vanish, that is, η' vanishes; *all* directions y' remain unchanged. This in turn means that *all* integral curves through each of the P_i remain fixed.

An arbitrary point P can always be joined to any of the P_i by an integral curve of the second order differential equation, and – being by this construction the intersection of invariant curves – it must also remain fixed: the transformation leaves *all* points fixed, so it is the identity ($\mathbf{X} = 0$). So starting from the assumption of (at least) nine linearly independent transformations $\mathbf{X}_A$, we have shown that by a suitable linear superposition we can construct $\mathbf{X} = a^A\mathbf{X}_A = 0$ from these transformations, which contradicts the assumption; no symmetry with more than eight parameters exists. Examples of an eight-parameter symmetry are known, so that the result is: *the maximum number of symmetries admitted by a second order ordinary differential equation is eight.*

To avoid the impression that all second order differential equations admit a symmetry, we close this section with a counterexample. It is the equation

$$y'' = xy + e^{y'} + e^{-y'}. \tag{4.30}$$

We leave the proof to the reader.

4.4 Higher order differential equations. The general *n*th order linear equation

For ordinary differential equations of order $n > 2$ the procedure for determining the symmetries is quite similar to that described in detail for the second order differential equations.

One first has to calculate the $\eta^{(i)}$, $i = 1, \ldots, n$, according to the definition (4.2), then to write down the symmetry condition in either of the forms (3.44) or (4.1) in full, and finally to determine the functions $\xi(x, y)$ and $\eta(x, y)$ from the resulting partial differential equations. With increasing n that will become a rather tiresome process, and computers may help to facilitate this. But although lengthy, the differential equations for ξ and η are always linear.

With some experience, symmetries can sometimes be detected simply by a careful inspection of the differential equation. The reason is that the symmetries that most frequently occur are of the forms $\partial/\partial x$; $x(\partial/\partial x)$; $\partial/\partial y$; $y(\partial/\partial y)$, and $ax(\partial/\partial x) + by(\partial/\partial y)$, which correspond to the finite transformations $\tilde{x} = x + \varepsilon$; $\tilde{x} = (1 + \varepsilon)x$; $\tilde{y} = y + \varepsilon$; $\tilde{y} = (1 + \varepsilon)y$, and $\tilde{x} = (1 + \varepsilon)^a x$, $\tilde{y} = (1 + \varepsilon)^b y$, respectively, and are easy to recognize. Examples are

$$2y'y''' - 3y''^2 = 0, \qquad \mathbf{X} = a\frac{\partial}{\partial x} + b\frac{\partial}{\partial y} + cy\frac{\partial}{\partial y}, \tag{4.31}$$

$$H(y^{iv}, y'', y', y) = 0, \qquad \mathbf{X} = \frac{\partial}{\partial x}, \tag{4.32}$$

$$y''' - yy'' + y'^2 = 0, \qquad \mathbf{X} = a\frac{\partial}{\partial x} + b\left(x\frac{\partial}{\partial x} - y\frac{\partial}{\partial y}\right). \tag{4.33}$$

Of course, these need not be the only symmetries!

As already stated in Section 4.1, the general linear differential equation

$$L(y) = \sum_{i=0}^{n} f_i(x)y^{(i)} = g(x) \tag{4.34}$$

is in some respects an exceptional case. It is known that its general solution can be written as

$$y(x) = y_{\mathrm{inh}}(x) + \sum_{k=1}^{n} c_k u_k(x), \tag{4.35}$$

where $y_{\mathrm{inh}}(x)$ denotes a special solution of the "inhomogeneous" equation (4.34) and the $u_k(x)$ are n linearly independent solutions of the homogeneous equation

$$L(y) = \sum_{i=0}^{n} f_i(x)y^{(i)} = 0. \tag{4.36}$$

That is to say, one can always add a constant multiple of any of the n linearly independent functions $u_k(x)$ to a solution and obtain another solution. But this means exactly that

$$\tilde{y} = y + \varepsilon \sum_{k=1}^{n} a^k u_k(x), \qquad a^k = \text{const.}, \tag{4.37}$$

is a symmetry transformation with generator

$$\mathbf{X} = \eta(x)\frac{\partial}{\partial y} = \sum_{k=1}^{n} a^k u_k(x)\frac{\partial}{\partial y}. \tag{4.38}$$

A linear differential equation of order n admits (at least) an n-parameter group of Lie point symmetries.

To begin with the negative aspect of this result, determining the symmetries (the function η) obviously means solving the (homogeneous) nth order differential equation since η is a solution to this equation. So the idea of using symmetries in solving a differential equation may not be of much use for linear equations – although we can state that the deeper reason for the tractability of linear equations is their high symmetry!

On the other hand, if one comes across a (nonlinear) differential equation with n or more symmetries, the existence of these symmetries may indicate that one is dealing with a linear differential equation in a disguised form, that is, in unsuitable coordinates. An example of this is provided by the differential equation (4.20), that is, by $y'' = (x - y)y'^3$, which admits an eight-parameter group of symmetries and can be transformed into a linear equation by simply exchanging x and y. So in the case of (at least) n symmetries it may be worthwhile to try to transform the given differential equation into an explicitly linear form, best done by trying to transform the generator into the form (4.38). But from the general point of view it is more important that the symmetry conditions are *always* linear, and if we are able to solve them, we may be able to construct the solution *without* transforming the differential equation into an explicitly linear form (if it is a transformed linear equation at all). We shall come back to this problem in Chapters 7 and 9 [for equation (4.20) in Section 7.5].

We close this section with a few remarks on the general structure of the set of solutions to the symmetry condition. We have already stressed that the symmetry condition is a *linear* differential equation for $\xi(x, y)$ and $\eta(x, y)$ but so far have not considered the implications of its being a *partial* differential equation. Solutions of partial differential equations may depend on arbitrary functions – is that to be expected here too?

To answer this question, we take the symmetry condition $\eta^{(n)} = \mathbf{X}\omega$ and look for those terms that contain nth partial derivatives of ξ or η. Because of

$$\begin{aligned}\eta^{(n)} &= \left(\frac{\partial}{\partial x} + y'\frac{\partial}{\partial y}\right)^n (\eta - y'\xi) + \cdots \\ &= \sum_{m=0}^{n} \binom{n}{m}\left[y'^m \frac{\partial^n \eta}{(\partial x)^{n-m}(\partial y)^m} - y'^{m+1}\frac{\partial^n \xi}{(\partial x)^{n-m}(\partial y)^m}\right] + \cdots,\end{aligned} \tag{4.39}$$

these terms can be picked out for $(n > 1)$ by equating to zero the coefficients of y'^k in the symmetry condition. The result is

$$\binom{n}{k}\frac{\partial^n \eta}{(\partial x)^{n-k}(\partial y)^k} - \binom{n}{k-1}\frac{\partial^n \xi}{(\partial x)^{n-k+1}(\partial y)^{k-1}} \tag{4.40}$$
$$= \text{terms linear in } \xi, \eta \text{ and their derivatives of lower order.}$$

These are $n + 2$ equations for the nth derivatives, and by differentiating them with respect to x or y, we obtain $2(n + 2)$ independent equations that enable us to give each of the $(n + 1)$ derivatives of ξ and η as a linear combination of the lower order derivatives. But such a system admits only a finite number of linearly independent solutions: at a given point (x, y), one can prescribe exactly ξ, η and all their lower order derivatives. That is, ξ and η will (for $n > 1$) depend on arbitrary constants but *not* on arbitrary functions!

Since in addition to (4.40) the rest of the symmetry condition has to be satisfied, the possible number of symmetries is much smaller than indicated by (4.40) alone. It can be shown that the number r of Lie point symmetries admitted by an nth order differential equation ($n > 2$) is restricted by $r \leqslant n+4$.

4.5 Exercises

1 Check that the differential equations (4.6)–(4.8) admit the respective symmetries!

2 Determine the symmetries of $y'' + y = 0$. Is there a transformation that transforms this differential equation into $y'' = 0$? [Try $\tilde{y} = yf(x)$, $\tilde{x} = g(x)$.]

3 Show that every linear homogeneous second order differential equation admits an eight-parameter group of symmetries by transforming it into $y'' = 0$. (Try to transform the solutions!)

4 Where does the proof for the nonexistence of a *nine*-parameter group of symmetries (for a second order differential equation) collapse if one tries to apply it to an *eight*-parameter group?

5 Show that (a) $y'' = xy + e^{y'} + e^{-y'}$ and (b) $y'' = \tan y'(y' \tan y - x^{-1})$ do not admit any Lie point symmetries.

6 Show that every first order differential equation $y' = \omega(x, y)$ can be transformed into $y' = 0$ by a point transformation.

7 Show that $y' = \psi(x)\varphi(y)$ admits the symmetry with generator $\mathbf{X} = \varphi\, \partial/\partial y$.

5

How to use Lie point symmetries: differential equations with one symmetry

5.1 First order differential equations

Suppose we have a first order differential equation

$$y' = \omega(x, y) \Leftrightarrow \mathbf{A}f = \left(\frac{\partial}{\partial x} + \omega\frac{\partial}{\partial y}\right)f = 0, \tag{5.1}$$

and we know that it admits a symmetry with generator

$$\mathbf{X} = \xi(x, y)\frac{\partial}{\partial x} + \eta(x, y)\frac{\partial}{\partial y}. \tag{5.2}$$

Can we use this knowledge to establish an integration procedure?

To answer this question, we consider a solution ϕ to $\mathbf{A}f = 0$. Since $\mathbf{X}$ is a symmetry, it maps solutions into solutions, and as *all* solutions to $\mathbf{A}f = 0$ are functions of ϕ, we have $\mathbf{X}\phi = \Omega(\phi)$. By introducing a suitable function $\varphi(\phi)$ by $\mathrm{d}\varphi/\mathrm{d}\phi = 1/\Omega(\phi)$, we can arrive at $\mathbf{X}\varphi = 1$. So the mere existence of a symmetry ensures that a solution φ to the system

$$\mathbf{X}\varphi = 1, \qquad \mathbf{A}\varphi = 0, \tag{5.3}$$

exists and, moreover, φ is uniquely determined by it; no further transformation $\tilde{\varphi} = \tilde{\varphi}(\varphi)$ is possible. Writing down these equations in full, we obtain

$$\begin{aligned} \mathbf{X}\varphi &= \xi\varphi_{,x} + \eta\phi_{,y} = 1, \\ \mathbf{A}\varphi &= \varphi_{,x} + \omega\varphi_{,y} = 0. \end{aligned} \tag{5.4}$$

If $\mathbf{X}$ is not a multiple of $\mathbf{A}$ [for a first order differential equation this may happen (compare Section 4.2), but then we would not call $\mathbf{X}$ a true symmetry!], the system (5.4) can be solved to give $\varphi_{,x}$ and $\varphi_{,y}$ according to

$$\varphi_{,x} = -\omega/(\eta - \xi\omega), \qquad \varphi_{,y} = 1/(\eta - \xi\omega). \tag{5.5}$$

In determining $\varphi(x, y)$, we need not worry about integrability conditions since the existence of a solution is ensured; we can simply go ahead and write

$$\varphi = \int \frac{\mathrm{d}y - \omega(x, y)\,\mathrm{d}x}{\eta(x, y) - \xi(x, y)\omega(x, y)}. \tag{5.6}$$

This is the (general) solution to our differential equation $\mathbf{A}f = 0$, and, according to the general scheme outlined in Section 3.2, φ is also a first integral of $y' = \omega(x, y)$. So we can state: *if a differential equation* $y' = \omega(x, y)$ *admits a (true) symmetry with generator* $\mathbf{X} = \xi(x, y)(\partial/\partial x) + \eta(x, y)(\partial/\partial y)$, *then its solution is given (in implicit form) via a line integral as*

$$\varphi(x, y) = \int \frac{\mathrm{d}y - \omega\,\mathrm{d}x}{\eta - \xi\omega} = \varphi_0 = \text{const.} \tag{5.7}$$

There is a close connection between this theorem and the well-known procedure for integrating the differential equation by means of an integrating factor. By definition, an integrating factor is a function $M(x, y)$ that makes $\mathrm{d}y - \omega(x, y)\mathrm{d}x$ the differential of a function φ,

$$\mathrm{d}\varphi = M(\mathrm{d}y - \omega\,\mathrm{d}x). \tag{5.8}$$

Comparing (5.7) and (5.8), we see that $(\eta - \xi\omega)^{-1}$ is an integrating factor, and the condition (4.5), which ensures that ξ, η are the components of a symmetry $\mathbf{X}$, is completely equivalent to the integrability condition $M_{,x} = -(M\omega)_{,y}$ of (5.8) with $M = (\eta - \xi\omega)^{-1}$. So we can restate the above theorem as: *if a differential equation* $y' = \omega(x, y)$ *admits a (true) symmetry* $\mathbf{X} = \xi(x, y)(\partial/\partial x) + \eta(x, y)(\partial/\partial y)$, *then* $M = (\eta - \xi\omega)^{-1}$ *is an integrating factor.*

Since neither a symmetry nor an integrating factor of a first order differential equation is in general obtainable by means of a straightforward algorithm, this theorem is more of theoretical than of practical value and only of rather limited importance in finding an integration procedure.

To give an application of the general theory, let us take the example (4.8) with $y' = 1 + \tan(x - y)/x$, $\xi = 0$, $\eta = [x\cos(x - y)]^{-1}$. Formula (5.6) yields

$$\begin{aligned} \varphi &= \int [x\cos(x - y)\mathrm{d}y - \{x\cos(x - y) + \sin(x - y)\}\mathrm{d}x] \\ &= -x\sin(x - y), \end{aligned} \tag{5.9}$$

and from $\varphi = \varphi_0 = \text{const}$ one obtains $y = x + \arcsin \varphi_0/x$.

For later use, we want to write formula (5.6) in a more formal way as

$$\varphi = \int \frac{\begin{vmatrix} \mathrm{d}\mathbf{r} \\ \mathbf{A} \end{vmatrix}}{\Delta} = \int \frac{\begin{vmatrix} \mathrm{d}x & \mathrm{d}y \\ 1 & \omega \end{vmatrix}}{\Delta}, \qquad \Delta = \begin{vmatrix} \mathbf{X} \\ \mathbf{A} \end{vmatrix} = \begin{vmatrix} \xi & \eta \\ 1 & \omega \end{vmatrix}. \tag{5.10}$$

5.2 Higher order differential equations

Suppose we are given an nth order differential equation as

$$y^{(n)} = \omega(x, y, y', \ldots, y^{(n-1)}), \tag{5.11}$$

or

$$\mathbf{A}f = \left(\frac{\partial}{\partial x} + y'\frac{\partial}{\partial y} + \cdots + \omega\frac{\partial}{\partial y^{(n-1)}}\right)f = 0, \tag{5.12}$$

and we know that it admits a symmetry

$$\mathbf{X} = \xi(x, y)\frac{\partial}{\partial x} + \eta(x, y)\frac{\partial}{\partial y} + \cdots + \eta^{(n-1)}\frac{\partial}{\partial y^{(n-1)}}. \tag{5.13}$$

Can this symmetry be used in an integration procedure, perhaps in a way similar to that used for a first order differential equation?

To analyse the problem, we assume that the symmetry generator $\mathbf{X}$ has been transformed to its normal form $\tilde{\mathbf{X}} = \partial/\partial s$ (compare Section 2.2). This can always be done by a transformation in (x, y)-space, for example,

$$\tilde{x} = t(x, y), \qquad \tilde{y} = s(x, y), \qquad \tilde{\mathbf{X}} = \frac{\partial}{\partial s}. \tag{5.14}$$

In these coordinates the differential equation (5.11) takes the form $s^{(n)} = \tilde{\omega}(t, s, s', \ldots, s^{(n-1)})$ for the function $s(t)$, but since the symmetry condition (3.39) now yields $\tilde{\mathbf{X}}\tilde{\omega} = \partial\tilde{\omega}/\partial s = 0$ (all $\tilde{\eta}^{(n)}$ being zero for $n \geqslant 1$), $\tilde{\omega}$ is independent of s:

$$s^{(n)} = \tilde{\omega}(t, s', s'', \ldots, s^{(n-1)}). \tag{5.15}$$

This is in fact a differential equation *of order* $n-1$ for the function s'! So if appropriate coordinates (t, s) have been found, the nth order differential equation has been reduced to one of order $n-1$ and a quadrature

$$s = \int s'(t)\,\mathrm{d}t. \tag{5.16}$$

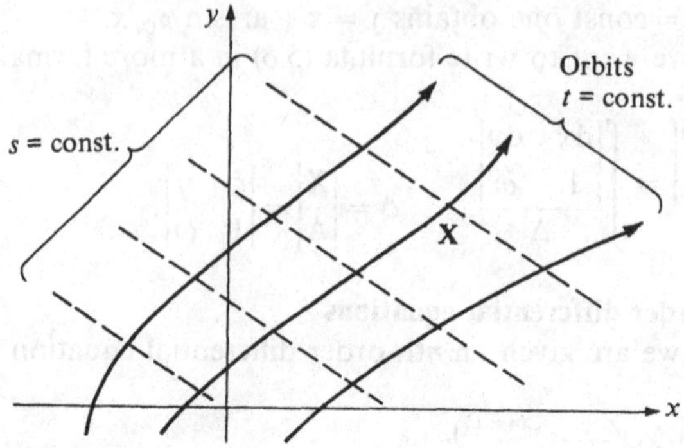

Figure 5.1. Coordinates $t(x, y)$ and $s(x, y)$ adjusted to a group with generator $\mathbf{X} = \xi\partial/\partial x + \eta\partial/\partial y = \partial/\partial s$.

How can one find the functions $s(x, y)$ and $t(x, y)$? To ensure the desired form $\partial/\partial s$ of the generator, they obviously have to satisfy

$$\mathbf{X}s = \xi(x, y)s_{,x} + \eta(x, y)s_{,y} = 1, \tag{5.17}$$

$$\mathbf{X}t = \xi(x, y)t_{,x} + \eta(x, y)t_{,y} = 0. \tag{5.18}$$

Geometrically, finding solutions $s(x, y)$ and $t(x, y)$ of (5.17) and (5.18) means introducing coordinates that have the orbits $t =$ const. of the group together with the lines $s =$ const. as coordinate lines (see Figure 5.1). Of course $s(x, y)$ is not uniquely fixed by (5.17): a transformation $\hat{s} = s + s_0(t)$ is always possible, that is, the origin of s can be chosen arbitrarily on each orbit. Nor is t determined by (5.18); a gauge change $\hat{t} = f(t)$ is always possible.

To solve (5.18), we first observe that on each orbit $t =$ const. the equation

$$\mathrm{d}t = 0 = t_{,x}\,\mathrm{d}x + t_{,y}\,\mathrm{d}y \tag{5.19}$$

holds. From this relation and equation (5.18) we can eliminate $t_{,x}$ and $t_{,y}$ and obtain the first order differential equation

$$\xi(x, y)\mathrm{d}y - \eta(x, y)\mathrm{d}x = 0, \tag{5.20}$$

t now being the constant of integration appearing in the solution to (5.20). The solution, for example, in the form $y = y(x, t)$, gives the orbits, each orbit labelled by its value of t, and by inversion we obtain $t = t(x, y)$.

To solve (5.17) (i.e., $\mathbf{X}s = 1$), we first introduce t as a new coordinate that we use together with, for example, y (if t does not depend on x, which happens for $\eta = 0$, we take t and x as coordinates). We then have

$$\mathbf{X} = (\mathbf{X}y)\frac{\partial}{\partial y} + (\mathbf{X}t)\frac{\partial}{\partial t} = (\mathbf{X}y)\frac{\partial}{\partial y} = \eta(y, t)\frac{\partial}{\partial y}. \tag{5.21}$$

The equation $\mathbf{X}s = 1$ now amounts to $\eta\, \partial s/\partial y = 1$, which is solved by

$$s(y, t) = \int \frac{\mathrm{d}y}{\eta(y, t)}, \qquad \eta \neq 0, \tag{5.22}$$

where t is to be kept constant when performing the integral. Having thus found $s(y, t)$, we have to replace t by $t(x, y)$ to obtain $s(x, y)$.

When $\eta = 0$, the relevant formula for s is, of course,

$$s(x, t) = \int \frac{\mathrm{d}x}{\xi(x, t)}, \qquad \xi \neq 0. \tag{5.23}$$

At first glance this procedure for getting the solutions $t(x, y)$ and $s(t, x)$ of $\mathbf{X}t = 0$, $\mathbf{X}s = 1$ may look complicated, but as the structure of the generator $\mathbf{X}$ is often simple, it can easily be carried through in most applications.

To summarize, *if we know that an nth order differential equation admits a Lie point symmetry, then we can reduce the problem to one first order differential equation* (5.20) (*to obtain the group orbits*), *to two quadratures* (5.16) *and* (5.22) *or* (5.23), *and to a differential equation* (5.15) *of order* $n - 1$.

A few examples will now illustrate the method.

If ω does not depend on y, the equation $y^{(n)} = \omega(x, y', \ldots, y^{(n-1)})$ already has the structure (5.15): it is already an equation of order $n - 1$ for y'.

If ω does not depend on x,

$$y^{(n)} = \omega(y, y', \ldots, y^{(n-1)}), \qquad \mathbf{X} = \frac{\partial}{\partial x}, \tag{5.24}$$

the simple change of coordinates

$$t = y, \qquad s = x, \qquad y' = \frac{\mathrm{d}y}{\mathrm{d}x} = \left(\frac{\mathrm{d}s}{\mathrm{d}t}\right)^{-1}, \qquad \mathbf{X} = \frac{\partial}{\partial s}, \quad \text{etc.,}$$

will do the job.

A case that frequently occurs is a differential equation that admits a symmetry of the form

$$\mathbf{X} = ax\frac{\partial}{\partial x} + by\frac{\partial}{\partial y}, \qquad ab \neq 0. \tag{5.25}$$

Differential equations that admit this symmetry have a scaling or similarity property: they do not change under the substitution $\tilde{x} = \lambda^a x$, $\tilde{y} = \lambda^b y$, $\tilde{y}' = \lambda^{b-a} y', \ldots$, a property that can easily be checked (compare the examples given in Section 4.4).

Here the differential equation of the orbits reads

$$\frac{\mathrm{d}y}{\mathrm{d}x} = \frac{by}{ax}, \tag{5.26}$$

which is solved by, for example, $y^a = tx^b$ and gives

$$t(x, y) = y^a/x^b \tag{5.27}$$

(or an arbitrary function of y^a/x^b). To determine $s(x, y)$, we must first use (5.22), which gives

$$s(y, t) = \int \frac{\mathrm{d}y}{by} = \ln(y^{1/b}) = s(y). \tag{5.28}$$

As remarked above, the function $s(x, y)$ is not uniquely defined; one can always add a function of t to it. If we choose to add $-(\ln t)/ab$, we arrive at $\hat{s} = (\ln x)/a$, which is of equal standing to (5.28).

If we apply these results to the differential equation (1.2), that is,

$$y'' = x^n y^2, \qquad \mathbf{X} = x\frac{\partial}{\partial x} - (n+2)y\frac{\partial}{\partial y} \tag{5.29}$$

(compare Table 4.1), we have, with $a = 1$ and $b = -(n+2)$,

$$t = yx^{n+2}, \qquad s = \ln x, \qquad x = e^s, \quad y = te^{-(n+2)s},$$

$$y' = \frac{\mathrm{d}y}{\mathrm{d}x} = e^{-(n+3)s}\left[\frac{\mathrm{d}t}{\mathrm{d}s} - (n+2)t\right], \tag{5.30}$$

and in the variables s, t the differential equation reads

$$s'' = -t^2 s'^3 + [(n+2)(n+3)ts' - (2n+5)]s'^2, \qquad s' \equiv \frac{\mathrm{d}s}{\mathrm{d}t}. \tag{5.31}$$

It does not look nice now (which means there are no obvious further symmetries), but it is of first order in s', as promised.

The fourth example is the differential equation (1.1) and one of its (many) symmetries given in (4.24):

$$y'' = (x-y)y'^3, \qquad \mathbf{X} = (x-y)\frac{\partial}{\partial x}. \tag{5.32}$$

From (5.20) and (5.23) we immediately obtain

$$t = y, \qquad s = \ln(x-y); \qquad y = t, \quad x = t + e^s, \tag{5.33}$$

which transforms the differential equation into

$$s'' + s'^2 + 1 = 0, \qquad s' = \frac{\mathrm{d}s}{\mathrm{d}t}, \quad \text{etc.,} \tag{5.34}$$

a differential equation that, by the way, can easily be solved (since there are additional symmetries).

The last example is concerned with a general linear homogeneous differential equation (compare Section 4.4). Suppose *one* solution $u(x)$ of

$$\sum_{i=0}^{n} f_i(x)y^{(i)} = 0 \tag{5.35}$$

is known. We then have to hand the symmetry $\mathbf{X} = u(x)\partial/\partial y$; see (4.38). What is the advice the symmetry considerations give us here? They say: introduce coordinates t, s with $\mathbf{X}t = 0$, $\mathbf{X}s = 1$; then you will arrive at a differential equation for $s' = \mathrm{d}s/\mathrm{d}t$ that is of order $n-1$. We immediately obtain from (5.18) and (5.22)

$$t = x, \qquad s = \int \frac{\mathrm{d}y}{u(t)} = \frac{y}{u(t)} = \frac{y}{u(x)}. \tag{5.36}$$

So the advice is: write y as a *product* of the known solution $u(x)$ and a function $s(x)$, which is exactly what all textbooks would tell us, of course without referring to any symmetry considerations.

Looking back at these examples, we can claim that the majority of classes of differential equations that can be integrated once (can be reduced in order) do have a symmetry, and the rules for treating them are special cases of the procedure based on the existence of a symmetry. Moreover, it is one of the most pleasant features of the symmetry approach that it does not rely on genius (of the user, not of the inventor!) but on the stubborn performance of a program. Even if we fail to recognize, for example, that a differential equation can be transformed into a linear one, the symmetry approach will give the correct prescription for treating it.

So far we have used the symmetry only in the integration procedure. But there is also a second type of application. Suppose you have with some luck found a special first integral φ, that is, a solution of $\mathbf{A}\varphi = 0$. Then, because of $[\mathbf{X}, \mathbf{A}] = \lambda \mathbf{A}$, $\phi = \mathbf{X}\varphi$ is also a first integral. It need not be a new one; it may happen to be trivial (constant) or functionally dependent on φ – but it is always worthwhile to try.

So from

$$y'' = -y, \quad \varphi = y'^2 + y^2,$$
$$\mathbf{X}_1 = \sin x \frac{\partial}{\partial y} - \cos x \frac{\partial}{\partial y'}, \qquad \mathbf{X}_2 = \frac{\partial}{\partial x}, \tag{5.37}$$

$\mathbf{X}_1 \varphi = 2(y \sin x - y' \cos x)$ yields a new first integral, but $\mathbf{X}_2 \varphi = 0$ is trivial.

We close with a remark on ordinary differential equations that are derivable from a Lagrangian $L = L(x, y, y', \ldots, y^{(m)})$, that is, differential equations that have the form

$$\frac{\partial L}{\partial y} + \sum_{i=1}^{m} \left(-\frac{\mathrm{d}}{\mathrm{d}x} \right)^i \frac{\partial L}{\partial y^{(i)}} = 0. \tag{5.38}$$

If such a differential equation admits a special Noether symmetry, that is, a symmetry $\mathbf{X}$ that leaves the action principle invariant,

$$\mathbf{X} \int L \,\mathrm{d}x = 0, \tag{5.39}$$

then this differential equation can be reduced in order by *two*. To get an impression of how this works, assume that the symmetry has (or has been transformed into) the form $\mathbf{X} = \partial/\partial y$, so that $\partial L/\partial y = 0$ holds. In that case equation (5.38) can be integrated once and yields

$$\sum_{i=1}^{m} \left(-\frac{\mathrm{d}}{\mathrm{d}x} \right)^{i-1} \frac{\partial L}{\partial y^{(i)}} = \text{const.} \tag{5.40}$$

But since L does not depend on y, this differential equation still admits the symmetry $\mathbf{X} = \partial/\partial y$, which can be used for a further reduction of order. We shall come back to this problem in Section 10.3.

Perhaps the reader wondered why we always used only *one* symmetry even if the existence of more symmetries was known. Is it possible, for example, to use two symmetries and reduce the order of the differential equation by two? To answer this question, we need some more knowledge about multiple-parameter groups and their generators, which will be provided in the following chapter. We shall then come back to this problem in Chapters 7–9.

5.3 Exercises

1 Integrate the first order differential equations (4.6) and (4.7) by using their respective symmetries.

2 Transform the symmetry (4.24) with $b = 0$ to coordinates t, s given in (5.33) and use the resulting generator to integrate (5.34).

3 How are the functions $s(x, y)$ obtained from (5.22) and (5.23) related to each other? Take $\mathbf{X} = x\,\partial/\partial x + y\,\partial/\partial y$ as an example!

4 Reduce the order of the differential equations (4.31) and (4.33).

6

Some basic properties of Lie algebras

6.1 The generators of multiple-parameter groups and their Lie algebras

If the group of symmetry transformations

$$\tilde{x} = \tilde{x}(x, y; \varepsilon_N), \qquad \tilde{y} = \tilde{y}(x, y; \varepsilon_N), \qquad N = 1, \ldots, r, \tag{6.1}$$

depends on more than one parameter ε_N, then the general infinitesimal generator of this group is a linear combination

$$\mathbf{X} = a^N \mathbf{X}_N, \qquad a^N = \text{const.}, \qquad N = 1, \ldots, r, \tag{6.2}$$

of r linearly independent basic generators $\mathbf{X}_N$, each of these $\mathbf{X}_N$ corresponding to one of the group parameters ε_N; compare Section 2.4. If we change the parametrization of the group, the result will be a linear transformation

$$\hat{\mathbf{X}}_N = B_N^M \mathbf{X}_M, \qquad B_N^M = \text{const.}, \qquad |B_N^M| \neq 0, \tag{6.3}$$

of the basic generators.

The main advantage in using the generators instead of the finite group transformations is that the generators are linear operators. That the superposition law (6.2) is also linear so that it does not matter in which order generators are added is just another way of saying that infinitesimal group transformations commute.

Finite transformations do not necessarily commute, as the example of three-dimensional rotations illustrates. How does this show up in the generators $\mathbf{X}_N$? Group theory shows that it is reflected in the commutators. The commutator of two generators $\mathbf{X}_N$ and $\mathbf{X}_M$ is defined by

$$[\mathbf{X}_N, \mathbf{X}_M] = \mathbf{X}_N\mathbf{X}_M - \mathbf{X}_M\mathbf{X}_N = -[\mathbf{X}_M, \mathbf{X}_N]. \tag{6.4}$$

It is again a linear operator if $\mathbf{X}_N$ and $\mathbf{X}_M$ are

$$\left[\xi_N\frac{\partial}{\partial x} + \eta_N\frac{\partial}{\partial y} + \cdots, \xi_M\frac{\partial}{\partial x} + \eta_M\frac{\partial}{\partial y} + \cdots\right]$$
$$= \Big((\mathbf{X}_N\xi_M) - (\mathbf{X}_M\xi_N)\Big)\frac{\partial}{\partial x} + \cdots. \tag{6.5}$$

To study the properties of commutators, we shall not start from the general abstract group theory but instead make use of the symmetry condition and definition.

For commutators like (6.4) the *Jacobi identity*

$$[\mathbf{X}_N, [\mathbf{X}_M, \mathbf{X}_P]] + [\mathbf{X}_M, [\mathbf{X}_P, \mathbf{X}_N]] + [\mathbf{X}_P, [\mathbf{X}_N, \mathbf{X}_M]] = 0 \tag{6.6}$$

always holds simply because of the definition of a commutator; the $\mathbf{X}_R$ need not be symmetries. We apply this identity to two symmetries $\mathbf{X}_1$ and $\mathbf{X}_2$ of a differential equation represented by the operator $\mathbf{A}$ and to $\mathbf{A}$ itself:

$$[\mathbf{X}_1, [\mathbf{X}_2, \mathbf{A}]] + [\mathbf{X}_2, [\mathbf{A}, \mathbf{X}_1]] + [\mathbf{A}, [\mathbf{X}_1, \mathbf{X}_2]] = 0. \tag{6.7}$$

Since $\mathbf{X}_1$ and $\mathbf{X}_2$ are symmetries,

$$[\mathbf{X}_1, \mathbf{A}] = \lambda_1\mathbf{A}, \qquad [\mathbf{X}_2, \mathbf{A}] = \lambda_2\mathbf{A}, \tag{6.8}$$

we obtain from (6.7)

$$\begin{aligned}[][[\mathbf{X}_1, \mathbf{X}_2], \mathbf{A}] &= [\mathbf{X}_1, \lambda_2\mathbf{A}] - [\mathbf{X}_2, \lambda_1\mathbf{A}] \\ &= \lambda_2\lambda_1\mathbf{A} + (\mathbf{X}_1\lambda_2)\mathbf{A} - \lambda_2\lambda_1\mathbf{A} - (\mathbf{X}_2\lambda_1)\mathbf{A} \\ &= \rho\mathbf{A};\end{aligned} \tag{6.9}$$

that is, the commutator $[\mathbf{X}_1, \mathbf{X}_2]$ is again a symmetry. By assumption, all symmetries can be written as a linear combination of the r basic generators, and so we have

$$[\mathbf{X}_1, \mathbf{X}_2] = c^N\mathbf{X}_N \tag{6.10}$$

or, more generally,

$$[\mathbf{X}_N, \mathbf{X}_M] = C^P_{NM}\mathbf{X}_P. \tag{6.11}$$

The constants C^P_{NM} are called the *structure constants* of the group. Because of the very definition of the commutator, they are antisymmetric in the two lower indices,

$$C^P_{NM} = -C^P_{MN}, \tag{6.12}$$

and because of the Jacobi identity (6.6), they have to satisfy the *Lie identity*

$$C^Q_{MP}C^R_{NQ} + C^Q_{PN}C^R_{MQ} + C^Q_{NM}C^R_{PQ} = 0. \tag{6.13}$$

The structure constants do not change under a coordinate transformation: if, in the notation of Section 2.2, we have

$$\mathbf{X}_1 = b^i_1\frac{\partial}{\partial x^i}, \qquad \mathbf{X}_2 = b^k_2\frac{\partial}{\partial x^k}, \tag{6.14}$$

then it follows from the transformation law (2.21), that is, from

$$x^{i'} = x^{i'}(x^i), \qquad b^{i'}_1 = b^i_1\frac{\partial x^{i'}}{\partial x^i}, \qquad b^{k'}_2 = \frac{\partial x^{k'}}{\partial x^k}b^k_2, \tag{6.15}$$

that the components $(\mathbf{X}_1 b^i_2) - (\mathbf{X}_2 b^i_1)$ of the commutator transform as the components of a vector; the structure constants are not affected.

The structure constants *do* change under a transformation of the basis. Inserting (6.3) into

$$[\hat{\mathbf{X}}_N, \hat{\mathbf{X}}_M] = \hat{C}^P_{NM}\hat{\mathbf{X}}_P, \tag{6.16}$$

we obtain the transformation law

$$B^R_N B^S_M C^Q_{RS} = B^Q_P \hat{C}^P_{NM}. \tag{6.17}$$

Transformations (6.3) of the basis can thus be used to simplify the structure constants of a given group.

The set of all $\{\mathbf{X}_N\}$, together with the commutators (structure constants), form the Lie algebra of the group under consideration. Conversely, it can be shown that for every set of constants C^P_{NM} that satisfy (6.12) and (6.13), a Lie group can locally uniquely be constructed.

6.2 Examples of Lie algebras

Before we come to study some examples, one point needs to be clarified. For the definition and discussion of generators and their commutators explicit use has been made of the differential operator **A**, which contains derivatives $\partial/\partial y^{(i)}$ up to some fixed order k, and thus the generators **X** are in fact generators $\mathbf{X}^{(k)}$ extended up to that order k. For a Lie point group, all components of the extended generators

$$\mathbf{X}_N^{(k)} = \xi_N \frac{\partial}{\partial x} + \eta_N \frac{\partial}{\partial y} + \eta'_N \frac{\partial}{\partial y'} + \cdots + \eta_N^{(k)} \frac{\partial}{\partial y^{(k)}},$$
$$\eta_N^{(k)} = \frac{\mathrm{d}\eta_N^{(k-1)}}{\mathrm{d}x} - y^{(k)} \frac{\mathrm{d}\xi_N}{\mathrm{d}x} \tag{6.18}$$

are completely determined by the generators

$$\mathbf{X}_N = \xi_N(x, y)\frac{\partial}{\partial x} + \eta_N(x, y)\frac{\partial}{\partial y} \tag{6.19}$$

in (x, y)-space. Do the commutators of the $\mathbf{X}_N$ also determine the commutators of their extensions; that is, does

$$[\mathbf{X}_N, \mathbf{X}_M] = C^P_{NM}\mathbf{X}_P \tag{6.20}$$

imply

$$[\mathbf{X}_N^{(k)}, \mathbf{X}_M^{(k)}] = C^P_{NM}\mathbf{X}_P^{(k)}, \tag{6.21}$$

with the same structure constants? Or do the structure constants depend on k? The answer is: they do not, and the proof (by induction) of this assertion will be given now.

Suppose (6.21) is valid for two arbitrary generators $\mathbf{X}_1^{(i)}$, $\mathbf{X}_2^{(i)}$ that are extended up to the ith derivative. We then show that the generators $\mathbf{X}_1^{(i+1)}$ and $\mathbf{X}_2^{(i+1)}$, which are extended one step further, also obey the same commutator relation. Since the commutator relations (6.20) do not change under a coordinate transformation, we can simplify the calculation by transforming $\mathbf{X}_1$ to its normal form $\mathbf{X}_1 = \partial/\partial x = \mathbf{X}_1^{(n)}$. The commutator relation $[\mathbf{X}_1^{(i)}, \mathbf{X}_2^{(i)}] = C^P_{12}\mathbf{X}_P^{(i)}$ is equivalent to

$$\frac{\partial}{\partial x}\xi_2 = C^P_{12}\xi_P, \qquad \frac{\partial}{\partial x}\eta_2^{(a)} = C^P_{12}\eta_P^{(a)}, \qquad 0 \leqslant a \leqslant i. \tag{6.22}$$

We must now compute

$$[\mathbf{X}_1^{(i+1)}, \mathbf{X}_2^{(i+1)}] = \left[\frac{\partial}{\partial x}, \mathbf{X}_2^{(i+1)}\right]. \tag{6.23}$$

The $(i+1)$th component of this commutator is

$$\frac{\partial}{\partial x}\eta_2^{(i+1)} = \frac{\partial}{\partial x}\left[\frac{\mathrm{d}\eta_2^{(i)}}{\mathrm{d}x} - y^{(i+1)}\frac{\mathrm{d}\xi_2}{\mathrm{d}x}\right], \tag{6.24}$$

where use has been made of the definition (6.18) of the $\eta_N^{(n)}$. The operators $\partial/\partial x$ and $\mathrm{d}/\mathrm{d}x = \partial/\partial x + y'\,\partial/\partial y + \cdots$ commute, and, using (6.22), we obtain

$$\begin{aligned}\frac{\partial}{\partial x}\eta_2^{(i+1)} &= \frac{\mathrm{d}}{\mathrm{d}x}\frac{\partial\eta_2^{(i)}}{\partial x} - y^{(i+1)}\frac{\mathrm{d}}{\mathrm{d}x}\frac{\partial\xi_2}{\partial x}\\ &= C_{12}^{P}\left(\frac{\mathrm{d}}{\mathrm{d}x}\eta_P^{(i)} - y^{(i+1)}\frac{\mathrm{d}\xi_P}{\mathrm{d}x}\right) = C_{12}^{P}\eta_P^{(i+1)}.\end{aligned} \tag{6.25}$$

If (6.22) is valid for $a \leqslant i$, then it is also valid for $a = i+1$: the $\mathbf{X}_1^{(i+1)}$ and $\mathbf{X}_2^{(i+1)}$ obey the same commutator relation as the $\mathbf{X}_1^{(i)}$ and $\mathbf{X}_2^{(i)}$. We have thus shown that the commutator of the extended generators equals the extension of the commutator of the nonextended generators. Or, in other words, the structure constants can be computed from the generators in x–y space; they do not depend on the order of extension.

We now give some examples of Lie algebras. The generators (2.53) of translations in the x and y directions commute:

$$\mathbf{X}_1 = \frac{\partial}{\partial x}, \qquad \mathbf{X}_2 = \frac{\partial}{\partial y}, \qquad [\mathbf{X}_1, \mathbf{X}_2] = 0. \tag{6.26}$$

Groups with vanishing structure constants are called *abelian groups.* Another example of vanishing structure constants (and thus an abelian group) are the generators (4.38) occurring as symmetries of linear differential equations,

$$\mathbf{X}_N = u_N(x)\frac{\partial}{\partial y}, \qquad [\mathbf{X}_N, \mathbf{X}_M] = 0. \tag{6.27}$$

For the symmetries of the equation $y'' = x^{-5}y^2$ (compare Table 4.1),we have

$$\begin{aligned}&\mathbf{X}_1 = x^2\frac{\partial}{\partial x} + xy\frac{\partial}{\partial y}, \qquad \mathbf{X}_2 = x\frac{\partial}{\partial x} + 3y\frac{\partial}{\partial y},\\ &[\mathbf{X}_1, \mathbf{X}_2] = (x^2 - 2x^2)\frac{\partial}{\partial x} + (3xy - xy - 3xy)\frac{\partial}{\partial y} = -\mathbf{X}_1.\end{aligned} \tag{6.28}$$

In (4.31) we found the group with generators

$$\mathbf{X}_1 = \frac{\partial}{\partial x}, \qquad \mathbf{X}_2 = \frac{\partial}{\partial y}, \qquad \mathbf{X}_3 = y\frac{\partial}{\partial y}. \tag{6.29}$$

Their commutators are

$$[\mathbf{X}_1, \mathbf{X}_2] = 0, \qquad [\mathbf{X}_1, \mathbf{X}_3] = 0, \qquad [\mathbf{X}_2, \mathbf{X}_3] = \mathbf{X}_2. \tag{6.30}$$

If we add a translation in the x direction and a rotation in the x–y plane to build up a Lie algebra,

$$\mathbf{X}_1 = \frac{\partial}{\partial x}, \qquad \mathbf{X}_2 = -y\frac{\partial}{\partial x} + x\frac{\partial}{\partial y}, \tag{6.31}$$

we obtain

$$[\mathbf{X}_1, \mathbf{X}_2] = \frac{\partial}{\partial y}. \tag{6.32}$$

The right side of this commutator *cannot* be written as a linear combination (with constant coefficients) of $\mathbf{X}_1$ and $\mathbf{X}_2$! This tells us that $\mathbf{X}_1$ and $\mathbf{X}_2$ alone are not the basis of a Lie algebra and we must add $\mathbf{X}_3 = \partial/\partial y$ to complete it:

$$[\mathbf{X}_1, \mathbf{X}_2] = \mathbf{X}_3, \qquad [\mathbf{X}_1, \mathbf{X}_3] = 0, \qquad [\mathbf{X}_2, \mathbf{X}_3] = \mathbf{X}_1. \tag{6.33}$$

Geometrically it is evident that a rotation generates a translation in the y direction from that in the x direction. In the context of symmetries of differential equations this example shows that if we have found (in some unsystematic way) several symmetries, then the commutators of these may lead to new symmetries. Of course, if one has *systematically* solved the symmetry conditions, then one has found *all* symmetries; nothing new can be gained from the commutators.

6.3 Subgroups and subalgebras

An r-parameter group G_r can have subgroups G_i, $0 \leqslant i \leqslant r$. A subgroup G_i consists of elements (transformations) of G_r that again form a group; that is, it contains the identity and for each transformation the inverse one, and the repeated application of its transformations is again an element of the subgroup. (In most cases the identity – $i = 0$ – and the group itself – $i = r$ – are of no interest and will be excluded.)

The generators $\mathbf{X}_\alpha$, $\alpha = 1, \ldots, i$, of a subgroup G_i form a Lie algebra $\{\mathbf{X}_\alpha\}$ that is a subalgebra of the Lie algebra $\{\mathbf{X}_N\}$, $N = 1, \ldots, r$, of the group G_r;

that is, the $\{\mathbf{X}_\alpha\}$ are a subset of the $\mathbf{X}_N$ with the property that their commutators can be expressed in terms of the elements of $\{\mathbf{X}_\alpha\}$ alone,

$$[\mathbf{X}_\alpha, \mathbf{X}_\beta] = C^{\nu}_{\alpha\beta}\mathbf{X}_\nu, \qquad \alpha, \beta, \nu = 1, \ldots, i. \tag{6.34}$$

If one simply takes i basis elements $\mathbf{X}_\alpha$ of a given Lie algebra, then in general they will *not* form a Lie algebra since we have

$$[\mathbf{X}_\alpha, \mathbf{X}_\beta] = C^{P}_{\alpha\beta}\mathbf{X}_P, \qquad \alpha, \beta = 1, \ldots, i < r, \qquad P = 1, \ldots, r; \tag{6.35}$$

the right side is not a linear combination of the $\mathbf{X}_\nu$ alone. But they will form a Lie algebra (and thus be the generators of a subgroup) if

$$C^{P}_{\alpha\beta} = 0 \quad \text{for } P > i \tag{6.36}$$

holds. Finding all subgroups G_i of a given group G_r means finding all subalgebras, that is, all sets $\{\mathbf{X}_\alpha\}$ of linear combinations $\mathbf{X}_\alpha = a^N_\alpha \mathbf{X}_N$ of the given Lie algebra basis vectors $\mathbf{X}_N$ that satisfy (6.34). Trivially all one-element sets $\mathbf{X}_1$ form a Lie algebra; no commutator condition need be satisfied.

For an abelian group all structure constants vanish, and so (6.34) is satisfied for an arbitrary choice of the $\mathbf{X}_\alpha$. An example is the symmetries (6.27) of a homogeneous linear differential equation.

The group (6.33) of motions in the x–y plane has the subgroup generated by the translations $\mathbf{X}_1$ and $\mathbf{X}_3$. Neither $\{\mathbf{X}_1, \mathbf{X}_2\}$ nor $\{\mathbf{X}_2, \mathbf{X}_3\}$ forms a subalgebra since the commutators yield $\mathbf{X}_3$ or $\mathbf{X}_1$, respectively.

The group with generators (6.29) has $\{\mathbf{X}_1, \mathbf{X}_2\}$, $\{\mathbf{X}_1, \mathbf{X}_3\}$, and $\{\mathbf{X}_2, \mathbf{X}_3\}$ as subalgebras.

A subalgebra that in principle can always be constructed is $\{[\mathbf{X}_N, \mathbf{X}_M]\}$: it consists of all linear combinations of the commutators (the commutators are in general not linearly independent, so they need not form a basis of this subalgebra). The group corresponding to this subalgebra is called the *derived group*. The derived group is the identity if and only if the group is abelian. The derived group can coincide with the group itself, as is the case for the algebra with three basis generators $\mathbf{X}_N$ that have the commutators

$$[\mathbf{X}_1, \mathbf{X}_2] = \mathbf{X}_3, \qquad [\mathbf{X}_2, \mathbf{X}_3] = \mathbf{X}_1, \qquad [\mathbf{X}_3, \mathbf{X}_1] = \mathbf{X}_2. \tag{6.37}$$

Here all three basic generators are (combinations of the) commutators.

An example where the derived group does not belong to one of these two extremal cases is provided by the motions (6.33) in the x–y plane. Here the derived group consists of the translations with generators $\mathbf{X}_1$ and $\mathbf{X}_3$.

If the derived group $G_{r'}$ does not coincide with the original group G_r, we can repeat the construction and consider the derived group $G_{r''}$ of the derived group $G_{r'}$. This procedure comes to an end when $G_{r^{(n+1)}} = G_{r^{(n)}}$. If $r^{(n)} = 0$ (i.e., if $G_{r^{(n)}}$ is the identity), then the group G_r is called *solvable* (or integrable).

According to this definition, every abelian group is solvable (with $r' = 0$). The group of motions (6.33) is also solvable with $r = 3$, $r' = 2$, and $r'' = 0$ (the commutator of $\mathbf{X}_1$ and $\mathbf{X}_3$ vanishes). The generators (6.37) belong to a group that is not solvable ($r = r' = 3$).

The concept of solvable (or unsolvable) groups becomes important when considering differential equations with more than two symmetries.

6.4 Realizations of Lie algebras. Invariants and differential invariants

When we determine the symmetries of a given differential equation, we find the generators explicitly in the form of (linear) differential operators $\mathbf{X}_N$, and we only afterwards compute the commutators to get the structure constants of the particular Lie algebra we have found. But we could also go backwards, that is, start from a given Lie algebra [a set of abstract elements $\mathbf{X}_N$ with a set of structure constants satisfying the conditions (6.12) and (6.13)] and ask which linear differential operators in (at most) two variables satisfy the given set of commutator relations with none of the operators $\mathbf{X}_N$ zero (i.e., we are asking for true realizations). We thus ask for possible realizations (or representations) of a Lie algebra. Note that if we prescribe the Lie algebra *and* the number of variables (the dimension of the space in which the differential operators $\mathbf{X}_N$ act), there need not be a solution; see the example given below.

If we start, for example, from equation (6.37), which defines a Lie algebra with three basic generators $\mathbf{X}_1$, $\mathbf{X}_2$, $\mathbf{X}_3$, we may ask whether there exist three- or two- or one-dimensional realizations. A three-dimensional realization is well known since (6.37) are exactly the commutator relations of the three-dimensional group of rotations; in (x, y, z)-space we have

$$\mathbf{X}_1 = -y\frac{\partial}{\partial x} + x\frac{\partial}{\partial y}, \qquad \mathbf{X}_2 = y\frac{\partial}{\partial z} - z\frac{\partial}{\partial y},$$

$$\mathbf{X}_3 = -z\frac{\partial}{\partial x} + x\frac{\partial}{\partial z}, \tag{6.38}$$

$$[\mathbf{X}_1, \mathbf{X}_2] = \mathbf{X}_3, \qquad [\mathbf{X}_2, \mathbf{X}_3] = \mathbf{X}_1, \qquad [\mathbf{X}_3, \mathbf{X}_1] = \mathbf{X}_2.$$

Geometry now tells us how to find a two-dimensional realization of this rotation group: one may take spherical coordinates (r, ϑ, φ) instead of (x, y, z). Then, because $\mathbf{X}_N r = \mathbf{X}_N(x^2 + y^2 + z^2)^{1/2} = 0$ for all three $\mathbf{X}_N$, the generators can be expressed in terms of $\partial/\partial\varphi$ and $\partial/\partial\vartheta$ (rotations can be expressed in terms of angles). Writing again x, y for φ, ϑ, we obtain as a two-dimensional representation

$$\mathbf{X}_1 = \frac{\partial}{\partial x}, \qquad \mathbf{X}_2 = \cot y \cos x \frac{\partial}{\partial x} + \sin x \frac{\partial}{\partial y},$$
$$\mathbf{X}_3 = -\sin x \cot y \frac{\partial}{\partial x} + \cos x \frac{\partial}{\partial y}. \tag{6.39}$$

There are no one-dimensional realizations of the rotation group. For if we transform, for example, $\mathbf{X}_1$ to its normal form, we should have

$$\mathbf{X}_1 = \frac{\partial}{\partial x}, \qquad \mathbf{X}_2 = f(x)\frac{\partial}{\partial x}, \tag{6.40}$$

and the commutator relations would give

$$\mathbf{X}_3 = f'(x)\frac{\partial}{\partial x}, \qquad f'' + f = 0, \qquad ff'' - f'^2 = 1, \tag{6.41}$$

but this is impossible for any (real) function $f(x)$.

In the context of symmetries of differential equations, a very important realization is that given by the extended generators in the space of solutions (first integrals) of the differential equation in question, that is, in the space of the n independent solutions φ^α of the differential equation

$$\mathbf{A}\varphi^\alpha = \left(\frac{\partial}{\partial x} + y'\frac{\partial}{\partial y} + \cdots + \omega\frac{\partial}{\partial y^{(n-1)}}\right)\varphi^\alpha(x, y, y', \ldots, y^{(n-1)}) = 0 \tag{6.42}$$

the realization being given by

$$\mathbf{X}_N = (\mathbf{X}_N\varphi^\alpha)\frac{\partial}{\partial\varphi^\alpha},$$
$$(\mathbf{X}_N\varphi^\alpha) \equiv \left(\xi_N\frac{\partial}{\partial x} + \eta_N\frac{\partial}{\partial y} + \eta'_N\frac{\partial}{\partial y'} + \cdots + \eta_N^{(n-1)}\frac{\partial}{\partial y^{(n-1)}}\right)\varphi^\alpha. \tag{6.43}$$

An interesting realization of the derived group $G_{r'}$ of a given group G_r makes direct use of the structure constants. It is defined by

$$\mathbf{X}_N = -C^P_{NA} z^A \frac{\partial}{\partial z^P} \tag{6.44}$$

and is called the *adjoint representation*. Note that although the indices $N, A, P, \ldots$ formally run from 1 to r, only r' of these operators are linearly independent since only r' of the $C^P_{NA}\mathbf{X}_P$ are assumed to be linearly independent (with constant coefficients). The proof that the operators (6.44) obey the correct commutator relations (6.11) is left to the reader.

For a given realization we may ask for the sets of equivalent points, that is, points that can be connected with each other by means of symmetry transformations. These sets are called *orbits* of the group. If the orbit coincides with the whole representation space, then the group acts transitively, or is *transitive*. If not, then the group acts *intransitively* and the space is divided into families of orbits. On each orbit, the group acts *simply* or *multiply transitively*, depending on whether two points can be connected by exactly one or by several different group transformations. The group of motions (rotation plus two translations) of the x–y plane is multiply transitive, for example, and the group (6.38) of three-dimensional rotations is intransitive, the two-dimensional orbits being the spheres $r = \text{const.}$ (and on each sphere the group is multiply transitive).

To decide whether a given group acts transitively and to find the dimension of the orbit(s), one has to determine the dimension of the space spanned (locally) by the generators $\mathbf{X}_N$ (which equals the dimension of the orbit). So, for example, the generators (6.29) $\mathbf{X}_1 = \partial/\partial x$, $\mathbf{X}_2 = \partial/\partial y$, and $\mathbf{X}_3 = y(\partial/\partial y)$ span a two-dimensional space, and the group is (multiply) transitive in (x, y)-space. The group of n generators $\mathbf{X}_N = u_N(x)\,\partial/\partial y$ is intransitive in (x, y)-space since all x-components of the generators are zero.

To find the orbits, one has to find the nonconstant solutions of

$$\mathbf{X}_N \phi^\alpha = 0 \quad \text{for all } N. \tag{6.45}$$

The surfaces (spaces, lines, etc.) $\phi^\alpha = \text{const.}$ then define the orbits. If no such solutions exist, the group acts transitively. So, for example, the translations $\mathbf{X}_1 = \partial/\partial x$, $\mathbf{X}_2 = \partial/\partial y$ in (x, y, z)-space have the orbits $\phi = z = \text{const.}$

The transitivity properties of a group depend heavily on the realization space. Rotations are intransitive in three-dimensional (x, y, z)-space but transitive on the spheres. Translations $\mathbf{X}_1 = \partial/\partial x$, $\mathbf{X}_2 = \partial/\partial y$, $[\mathbf{X}_1, \mathbf{X}_2] = 0$, are transitive in (x, y)-space, but their extensions to the (x, y, y')-space are not. More generally, the extensions of a point transformation have transitivity properties different from those of the point transformations, and by extending to derivatives of order $r + 1$, we can find an intransitive realization for any point group G_r. Or, stated differently, for a given group of point transformations we can always find solutions of (6.45), that is, of

$$\mathbf{X}^{(k)}_N \phi^\alpha(x, y, y', \ldots, y^{(k)}) = 0, \qquad N = 1, \ldots, r, \tag{6.46}$$

if we choose a large enough k. Functions that are solutions to (6.46), that is, functions that do not change under the action of (the generators of) a group, are called *invariants.* If they depend on the derivatives, and to stress this point, they are called *differential invariants.* The *order* of such an invariant is the order of the highest derivative appearing in it.

How does the number of differential invariants change with the order k? Equation (6.46), from which the invariants are to be determined, is in fact a system of r linear differential equations in the $k+2$ variables $\{x, y, y', \ldots, y^{(k)}\}$. Since it can be shown that the integrability conditions of such systems say that all commutators between the $\mathbf{X}_N$ must be linear combinations of the $\mathbf{X}_N$ and since this is guaranteed by the group properties, equations (6.46) can be treated as an algebraic system:

$$\begin{aligned} \xi_1 \phi_{,x} + \eta_1 \phi_{,y} + \cdots + \eta_1^{(k)} \phi_{,y^{(k)}} &= 0, \\ \vdots \qquad \vdots \qquad\qquad \vdots \qquad & \\ \xi_r \phi_{,x} + \eta_r \phi_{,y} + \cdots + \eta_r^{(k)} \phi_{,y^{(k)}} &= 0, \end{aligned} \tag{6.47}$$

from which to determine the derivatives of ϕ. It is well known that nontrivial solutions to (6.47) will exist only if the rank m of the $r \times (k+2)$ matrix of its coefficients (the components of the $\mathbf{X}_N^{(k)}$) is less than $k+2$ and the number of independent solutions is $k+2-m$. So if we enlarge k by one, we have to add one column to that matrix, which may or may not enlarge the rank m by one, but the number of independent solutions will grow at most by one.

This result can be used to construct all differential invariants of higher order given only the first two in that hierarchy. The procedure runs as follows. Take these first two differential invariants φ and ψ, one of which is of order, say, i, and construct

$$\rho = \frac{\mathrm{d}\varphi}{\mathrm{d}\psi} = \frac{\mathrm{d}\varphi/\mathrm{d}x}{\mathrm{d}\psi/\mathrm{d}x}. \tag{6.48}$$

Then ρ is also an invariant and is of order $i+1$, and hence the (only) invariant of order $i+1$ is known. Then take $\mathrm{d}^2\varphi/\mathrm{d}\psi^2, \mathrm{d}^3\varphi/\mathrm{d}\psi^3, \ldots$ to obtain all higher order invariants of the given group of point symmetries.

To prove the above assertion, we must show that $\mathbf{X}_N\varphi = 0 = \mathbf{X}_N\psi$ implies $\mathbf{X}_N\rho = 0$. Because of

$$\begin{aligned} \left[\frac{\partial}{\partial y^{(p)}}, \frac{\mathrm{d}}{\mathrm{d}x}\right] &= \left[\frac{\partial}{\partial y^{(p)}}, \frac{\partial}{\partial x} + y'\frac{\partial}{\partial y} + \cdots + y^{(k+1)}\frac{\partial}{\partial y^{(k)}}\right] \\ &= \frac{\partial}{\partial y^{(p-1)}}, \end{aligned} \tag{6.49}$$

we have

$$\left[\mathbf{X}_N, \frac{\mathrm{d}}{\mathrm{d}x}\right] = -\frac{\mathrm{d}\xi_N}{\mathrm{d}x}\frac{\mathrm{d}}{\mathrm{d}x}, \tag{6.50}$$

and if we apply this to (6.48), we obtain indeed

$$\begin{aligned} \mathbf{X}_N\rho &= \mathbf{X}_N\frac{\mathrm{d}\varphi/\mathrm{d}x}{\mathrm{d}\psi/\mathrm{d}x} \\ &= \left(\frac{\mathrm{d}\psi}{\mathrm{d}x}\right)^{-2}\left[\frac{\mathrm{d}\psi}{\mathrm{d}x}\left(\frac{\mathrm{d}}{\mathrm{d}x}(\mathbf{X}_N\varphi) - \frac{\mathrm{d}\xi_N}{\mathrm{d}x}\frac{\mathrm{d}\varphi}{\mathrm{d}x}\right)\right. \\ &\quad \left. - \frac{\mathrm{d}\varphi}{\mathrm{d}x}\left(\frac{\mathrm{d}}{\mathrm{d}x}(\mathbf{X}_N\psi) - \frac{\mathrm{d}\xi_N}{\mathrm{d}x}\frac{\mathrm{d}\psi}{\mathrm{d}x}\right)\right] \\ &= 0. \end{aligned} \tag{6.51}$$

We close with a remark on how to check the transitivity of a given n-parameter group (with generators $\mathbf{X}_N$) in the n-dimensional space of first integrals of an nth order differential equation. To act transitively in that space, the group must not admit invariants; that is, solutions ϕ to the system

$$\begin{aligned} \mathbf{A}\phi &= \phi_{,x} + y'\phi_{,y} + \cdots + \omega\phi_{,y^{(n-1)}} = 0, \\ \mathbf{X}_N\phi &= \xi_N\phi_{,x} + \eta_N\phi_{,y} + \cdots + \eta_N^{(n-1)}\phi_{,y^{(n-1)}} = 0, \quad N = 1, \ldots, n, \end{aligned} \tag{6.52}$$

must not exist. As done above in the case of the system (6.47), we can treat (6.52) as an algebraic system and conclude that its determinant must be nonzero: a Lie algebra of dimension n acts transitively in the space of first integrals of an nth order differential equation if

$$\left\| \begin{matrix} \mathbf{A} \\ \mathbf{X}_N \end{matrix} \right\| \neq 0 \tag{6.53}$$

holds.

6.5 Nth order differential equations with multiple-parameter symmetry groups: an outlook

What have all these details of Lie algebras, their realizations, and their invariants to do with the integration of differential equations known to admit symmetries? It will turn out that the advantage we can obtain from the existence of a group of symmetries will depend on the order of that group (compared with the order of the differential equation) and on the type of the algebra. The details of the integration procedure will also depend on the transitivity properties in the solution space, and we shall benefit from the use of invariants and differential invariants at nearly every step (although we shall not always stress this fact).

To cover all possible cases of Lie algebras and differential equations, one could proceed as follows: (1) Prescribe the order n of the differential equation starting with $n = 2$. (2) Find all possible Lie algebras with $r \leqslant n + 4$ ($r \leqslant 8$ for $n = 2$) generators that have realizations in two variables, that is, that are Lie point symmetries and could appear as symmetries of differential equations. (3) Find suitable integration techniques for each case. Note that point 2 of this program is a purely algebraic task; it can be (and has been) carried through without any reference to differential equations.

Although simple in structure, this is in fact an infinite program. So we must select some topics, which will be the second order differential equations and some general remarks on how to deal with higher order equations. These will be the subjects of the following chapters.

6.6 Exercises

1 Show that the eight generators of the projective transformation (4.26) form a Lie algebra (without computing *all* commutators).

2 Show that the group G_3 with commutators $[\mathbf{X}_1, \mathbf{X}_2] = \mathbf{X}_1$, $[\mathbf{X}_2, \mathbf{X}_3] = \mathbf{X}_3$, and $[\mathbf{X}_1, \mathbf{X}_3] = 2\mathbf{X}_2$ has a one-dimensional realization.

3 Show that in the realization (6.43) of symmetries in the space of solutions none of the $\mathbf{X}_N$ can vanish.

4 Show that the $\mathbf{X}_N$ as given by (6.44) obey the correct commutator relations.

5 How does the realization (6.44) read for the group of three-dimensional rotations and for the group of Exercise 6.2?

6 Show that the generators $\{[\mathbf{X}_N, \mathbf{X}_M]\}$ of the derived group indeed form a Lie algebra.

7 Show that every G_2 (group with two generators) is solvable.

8 Show that the generators (4.38) of symmetries of an nth order linear differential equation are transitive in the space of solutions.

9 Why are the integrability conditions of the system (6.52) satisfied?

7

How to use Lie point symmetries: second order differential equations admitting a G_2

7.1 A classification of the possible subcases, and ways one might proceed

If a second order differential equation

$$y'' = \omega(x, y, y') \Leftrightarrow \mathbf{A}f = \left(\frac{\partial}{\partial x} + y'\frac{\partial}{\partial y} + \omega\frac{\partial}{\partial y'}\right)f(x, y, y') = 0 \tag{7.1}$$

admits a two-parameter group G_2, then there exist two generators $\mathbf{X}_1$ and $\mathbf{X}_2$ that form a Lie algebra; that is, their commutator is a linear combination of these two generators,

$$[\mathbf{X}_1, \mathbf{X}_2] = C^p_{12}\mathbf{X}_p = c_1\mathbf{X}_1 + c_2\mathbf{X}_2. \tag{7.2}$$

Two cases have to be distinguished. Either both structure constants c_1 and c_2 are zero and the group is an abelian one, sometimes designated G_2I,

$$G_2\text{I}: \; [\mathbf{X}_1, \mathbf{X}_2] = 0, \tag{7.3}$$

or at least one of the structure constants, say, c_1, does not vanish. We can then make a transformation (6.3) of the basis to simplify the structure constants, for example, by $\hat{\mathbf{X}}_1 = c_1\mathbf{X}_1 + c_2\mathbf{X}_2$, $\hat{\mathbf{X}}_2 = \mathbf{X}_2/c_1$, to have $[\hat{\mathbf{X}}_1, \hat{\mathbf{X}}_2] = \hat{\mathbf{X}}_1$. That is, we can always assume, in the noncommuting case (designated G_2II),

$$G_2\text{II: } [\mathbf{X}_1, \mathbf{X}_2] = \mathbf{X}_1. \tag{7.4}$$

To summarize, there are two distinct types of two-parameter groups G_2, which are given by (7.3) and (7.4).

We must now find a method (for each case) of using the two symmetries in integrating the given differential equation. It seems to be a good proposal to generalize the idea that proved successful in the case of one symmetry, namely, to transform the symmetry generators to some simple normal forms and then to see whether we can deal with the differential equation in these new coordinates. We shall pursue this idea now.

We start with the case of two commuting generators. We can always transform $\mathbf{X}_1$ to its normal form $\mathbf{X}_1 = \partial/\partial s$ by introducing coordinates $s(x, y)$, $t(x, y)$. The general form of a second generator $\mathbf{X}_2$ in these coordinates is $\mathbf{X}_2 = a(s, t)\,\partial/\partial s + b(s, t)\,\partial/\partial t$, but to give $[\mathbf{X}_1, \mathbf{X}_2] = 0$, it must be

$$\mathbf{X}_1 = \partial/\partial s, \quad \mathbf{X}_2 = a(t)\,\partial/\partial s + b(t)\,\partial/\partial t. \tag{7.5}$$

The transformations that leave $\mathbf{X}_1$ invariant (but change $\mathbf{X}_2$) are

$$u = s + h(t), \qquad v = v(t). \tag{7.6}$$

They give

$$\mathbf{X}_1 = \frac{\partial}{\partial u},$$

$$\mathbf{X}_2 = (\mathbf{X}_2 u)\frac{\partial}{\partial u} + (\mathbf{X}_2 v)\frac{\partial}{\partial v} = (a + bh')\frac{\partial}{\partial u} + bv'\frac{\partial}{\partial v}. \tag{7.7}$$

For $b = 0$, we choose $v = a$ and have

$$\mathbf{X}_1 = \frac{\partial}{\partial u}, \quad \mathbf{X}_2 = v\frac{\partial}{\partial u}. \tag{7.8}$$

For $b \neq 0$, we choose $v' = 1/b$ and $h' = -a/b$ and obtain

$$\mathbf{X}_1 = \frac{\partial}{\partial u}, \quad \mathbf{X}_2 = \frac{\partial}{\partial v}. \tag{7.9}$$

To get the second order differential equations that admit (7.8) or (7.9) as

Table 7.1. *Normal forms of the generators of a G_2 and the respective forms of the differential equation for $s = s(t)$, with* $\delta = \begin{vmatrix} \xi_1 \eta_1 \\ \xi_2 \eta_2 \end{vmatrix}$

$G_2\mathrm{I}$: $[\mathbf{X}_1, \mathbf{X}_2] = 0$	Ia, $\delta \neq 0$	$\mathbf{X}_1 = \frac{\partial}{\partial s}, \mathbf{X}_2 = \frac{\partial}{\partial t}$	$s'' = \hat{\omega}(s')$
	Ib, $\delta = 0$	$\mathbf{X}_1 = \frac{\partial}{\partial s}, \mathbf{X}_2 = t\frac{\partial}{\partial s}$	$s'' = \hat{\omega}(t)$
$G_2\mathrm{II}$: $[\mathbf{X}_1, \mathbf{X}_2] = \mathbf{X}_1$	IIa, $\delta \neq 0$	$\mathbf{X}_1 = \frac{\partial}{\partial s}, \mathbf{X}_2 = t\frac{\partial}{\partial t} + s\frac{\partial}{\partial s}$	$s'' = \frac{\hat{\omega}(s')}{t}$
	IIb, $\delta = 0$	$\mathbf{X}_1 = \frac{\partial}{\partial s}, \mathbf{X}_2 = s\frac{\partial}{\partial s}$	$s'' = s'\hat{\omega}(t)$

symmetries, we have to decide whether to take v or u as the independent variable. If in the case (7.8) we choose to write $v''(u) = \omega(u, v, v')$, then the symmetry conditions $\mathbf{X}_N \omega = \eta''_N \pmod{v'' = \omega}$ amount to $\omega_{,u} = 0$ and $v'\omega_{,v'} = 3\omega$; that is, the differential equation must have the form $v'' = v''^3 \hat{\omega}(v)$. If, on the other hand, we choose to write $u''(v) = \tilde{\omega}(v, u, u')$, then the symmetry conditions $\mathbf{X}_N \tilde{\omega} = \eta''_N \pmod{u'' = \tilde{\omega}}$ lead to $\tilde{\omega}_{,u} = 0 = \tilde{\omega}_{,u'}$, that is, to $u'' = \tilde{\omega}(v)$, which is even simpler.

Making the same analysis for the case (7.9) and going through the same routine for noncommuting generators, we arrive at Table 7.1, in which the symbols t and s have (again) been used after performing all necessary transformations. By the way, the possible forms of the generators (for each type of commutators) differ in whether they are or are not transitive in (x, y) space, which shows up in whether δ vanishes or not.

We see that in all four cases the resulting differential equations can easily be solved in terms of quadratures. So a possible strategy for dealing with second order differential equations admitting a G_2 of Lie point symmetries is to first transform the generators to their normal forms and then integrate the resulting differential equations. But we shall see that there is also a different way to use the symmetries, the main idea being to transform the generators to some normal forms not in their realization in the space of the dependent and independent variables but in their realization in the space of first integrals. The details of the two approaches will be presented now.

7.2 The first integration strategy: normal forms of generators in the space of variables

If we have two symmetries $\mathbf{X}_1$ and $\mathbf{X}_2$, we must check whether they commute or not (type I or II of Table 7.1) and whether they are linearly independent or not in the space of dependent and independent variables (x, y) or (s, t) (type a or b of Table 7.1). We then know which normal form we have to achieve. The integration procedure is a little bit different in the four cases that arise. In each case, we first have to determine t and s and then integrate the resulting simple differential equation for $s(t)$.

Case Ia. This case is characterized invariantly by

$$[\mathbf{X}_1, \mathbf{X}_2] = 0, \qquad \delta = \begin{vmatrix} \mathbf{X}_1 \\ \mathbf{X}_2 \end{vmatrix} = \begin{vmatrix} \xi_1 & \eta_1 \\ \xi_2 & \eta_2 \end{vmatrix} \neq 0. \tag{7.10}$$

We know that functions $t(x, y)$ and $s(x, y)$ exist that satisfy

$$\mathbf{X}_1 t = \xi_1(x, y)t_{,x} + \eta_1(x, y)t_{,y} = 0, \tag{7.11}$$
$$\mathbf{X}_2 t = \xi_2(x, y)t_{,x} + \eta_2(x, y)t_{,y} = 1,$$

and

$$\mathbf{X}_1 s = \xi_1 s_{,x} + \eta_1 s_{,y} = 1, \tag{7.12}$$
$$\mathbf{X}_2 s = \xi_2 s_{,x} + \eta_2 s_{,y} = 0.$$

Since the existence of solutions is ensured, we need not bother about integrability conditions but can simply compute the derivatives of t and s from these equations (by assumption, the determinant $\xi_1\eta_2 - \xi_2\eta_1$ does not vanish!) and obtain t and s as line integrals,

$$t(x, y) = \int (t_{,x}\,\mathrm{d}x + t_{,y}\,\mathrm{d}y) = \int \frac{-\eta_1\,\mathrm{d}x + \xi_1\,\mathrm{d}y}{\xi_1\eta_2 - \xi_2\eta_1}, \tag{7.13}$$

$$s(x, y) = \int (s_{,x}\,\mathrm{d}x + s_{,y}\,\mathrm{d}y) = \int \frac{\eta_2\,\mathrm{d}x - \xi_2\,\mathrm{d}y}{\xi_1\eta_2 - \xi_2\eta_1}. \tag{7.14}$$

We now transform the differential equation $y'' = \omega(x, y, y')$. We know already from Table 7.1 that it will be of the form

$$s''(t) = \hat{\omega}(s'), \tag{7.15}$$

but we still have to determine the function $\hat{\omega}$. To obtain $\hat{\omega}$, we first use (7.13) and (7.14) to calculate s',

$$s' = \frac{ds}{dt} = \frac{s_{,x}\,dx + s_{,y}\,dy}{t_{,x}\,dx + t_{,y}\,dy} = \frac{\eta_2\,dx - \xi_2\,dy}{-\eta_1\,dx + \xi_1\,dy} = \frac{y'\xi_2 - \eta_2}{\eta_1 - y'\xi_1}, \tag{7.16}$$

which gives us

$$y' = \frac{\eta_2 + s'\eta_1}{\xi_2 + s'\xi_1}, \tag{7.17}$$

and in a similar way we get y'' from ds'/dt. After an elimination procedure we then arrive at (7.15). This simple differential equation can be integrated to give

$$t + \varphi_0 = \int \frac{ds'}{\hat{\omega}(s')} \tag{7.18}$$

and, with $s' = f(t + \varphi_0)$,

$$s(t, \varphi_0, \psi_0) = \int f(t + \varphi_0)\,dt + \psi_0. \tag{7.19}$$

From $s(x, y)$, $t(x, y)$, and $s(t, \varphi_0, \psi_0)$ we can eliminate s and t and obtain the solution $y = y(x, \varphi_0, \psi_0)$.

Case Ib. This is characterized by

$$[\mathbf{X}_1, \mathbf{X}_2] = 0, \qquad \delta = \begin{vmatrix} \mathbf{X}_1 \\ \mathbf{X}_2 \end{vmatrix} = \begin{vmatrix} \xi_1 & \eta_1 \\ \xi_2 & \eta_2 \end{vmatrix} = 0. \tag{7.20}$$

So if $\xi_1\eta_2 - \xi_2\eta_1$ vanishes, we know from Table 7.1 that $\mathbf{X}_2 = t(x, y)\mathbf{X}_1$ holds, from which we can immediately read off the function $t(x, y)$. To get the function s, we first introduce t as a new variable instead of x (or instead of y if $t_{,x} = 0$). We then have

$$\mathbf{X}_1 s(y, t) = (\mathbf{X}_1 y)s_{,y} + (\mathbf{X}_1 t)s_{,t} = \eta_1(y, t)s_{,y} = 1, \tag{7.21}$$

that is,

$$s(y, t) = \int \frac{dy}{\eta_1(y, t)}. \tag{7.22}$$

The transformation of the differential equation leads to $s''(t) = \hat{\omega}(t)$, which is integrated by

$$s(t) = \iint^t \omega(t')\,dt'\,dt + \varphi_0 t + \psi_0. \tag{7.23}$$

We can eliminate s from (7.22) and (7.23) to get $t = f(y, \varphi_0, \psi_0)$, which together with $t = t(x, y)$ yields $y = y(x, \varphi_0, \psi_0)$.

Case IIa. This is characterized by

$$[\mathbf{X}_1, \mathbf{X}_2] = \mathbf{X}_1, \qquad \delta = \begin{vmatrix} \mathbf{X}_1 \\ \mathbf{X}_2 \end{vmatrix} = \begin{vmatrix} \xi_1 & \eta_1 \\ \xi_2 & \eta_2 \end{vmatrix} \neq 0. \tag{7.24}$$

To determine $t(x, y)$, we write $u = \ln t$ and have from Table 7.1

$$\begin{aligned} \mathbf{X}_1 u(x, y) &= \xi_1 u_{,x} + \eta_1 u_{,y} = 0, \\ \mathbf{X}_2 u(x, y) &= \xi_2 u_{,x} + \eta_2 u_{,y} = 1. \end{aligned} \tag{7.25}$$

The solution of (7.25) is obviously given by the line integral

$$u(x, y) = \ln t(x, y) = \int \frac{\eta_1\,dx - \xi_1 dy}{\xi_1\eta_2 - \xi_2\eta_1}. \tag{7.26}$$

We now use t as a coordinate instead of x (or instead of y if $t_{,x} = 0$) and make the *ansatz* $s = t\cdot v(y, t)$ for s. The function v then (because $\mathbf{X}_1 s = 1$, $\mathbf{X}_2 s = s$) has to satisfy

$$\begin{aligned} t\mathbf{X}_1 v &= t\eta_1(y, t) v_{,y} = 1, \\ \mathbf{X}_2 v &= t v_{,t} + \eta_2(y, t) v_{,y} = 0. \end{aligned} \tag{7.27}$$

The solution of (7.27) is

$$s(y, t) = t v(y, t) = t \int \left(\frac{dy}{t\eta_1} - \frac{\eta_2 dt}{t^2 \eta_1} \right). \tag{7.28}$$

In variables s, t the differential equation $y'' = \omega(x, y, y')$ then takes the form $s'' = \hat{\omega}(s')/t$, which can easily be solved by

$$\int \frac{ds'}{\hat{\omega}(s')} = \ln t + \varphi_0 \tag{7.29}$$

and, with $s' = g(t, \varphi_0)$,

$$s(t) = \int g(t, \varphi_0)\,\mathrm{d}t + \psi_0. \tag{7.30}$$

From (7.26), (7.28), and (7.30) we can eliminate s and t to obtain $y = y(x, \varphi_0, \psi_0)$.

Case IIb. Finally, this is characterized by

$$[\mathbf{X}_1, \mathbf{X}_2] = \mathbf{X}_1, \qquad \delta = \begin{vmatrix} \mathbf{X}_1 \\ \mathbf{X}_2 \end{vmatrix} = \begin{vmatrix} \xi_1 & \eta_1 \\ \xi_2 & \eta_2 \end{vmatrix} = 0. \tag{7.31}$$

As there is a linear relation $\mathbf{X}_2 = s(x, y)\mathbf{X}_1$, we immediately know the function $s(x, y)$. To obtain $t(x, y)$, the equation

$$\mathbf{X}_1 t = \xi_1 t_{,x} + \eta_1 t_{,y} = \xi_1 \left(\frac{\partial}{\partial x} + \frac{\eta_1}{\xi_1} \frac{\partial}{\partial y} \right) t = 0 \tag{7.32}$$

needs to be solved. This partial differential equation is equivalent to the first order equation for $y = y(x, t_0)$,

$$y' = \eta_1 / \xi_1. \tag{7.33}$$

No extra trick is available to solve it; it must be done by hand [unless $\xi_1 \eta_1 = 0$ and (7.32) is trivial]. If we have $y = y(x, t_0)$, invert it to get $t = t_0(x, y)$.

In coordinates s, t the differential equation $y'' = \omega(x, y, y')$ reads $s'' = s'\hat{\omega}(t)$, which is solved by

$$s(t) = \int \varphi_0 \left(\exp \int^t \hat{\omega}(t')\,\mathrm{d}t' \right) \mathrm{d}t + \psi_0. \tag{7.34}$$

From $t = t(x, y)$, $s = s(x, y)$, and $s = s(t, \varphi_0, \psi_0)$ one has to eliminate s and t to obtain $y(x, \varphi_0, \psi_0)$.

Looking back at these four cases, we see that typically we encounter systems of two partial differential equations like (7.11), (7.12), or (7.25), which can be solved by line integrals. Since the existence of a solution is ensured from the very beginning by the assumption of a symmetry group, no integrability conditions need to be considered. In all cases except case IIb the normal forms of the generators can be obtained by performing those line integrals and some elimination and transformation of variables and we are never told "Try to solve this differential equation!" In variables s and t the original differential equation is so simple that it can be solved by quadratures.

7.3 The second integration strategy: normal forms of generators in the space of first integrals

The differential equation to be solved,

$$\mathbf{A}f = \left(\frac{\partial}{\partial x} + y'\frac{\partial}{\partial y} + \omega(x, y, y')\frac{\partial}{\partial y'}\right)f = 0, \tag{7.35}$$

has two functionally independent solutions $\varphi(x, y, y')$ and $\psi(x, y, y')$. If we use φ and ψ and a third independent variable χ (which, e.g., satisfies $\mathbf{A}\chi = 1$) as variables instead of x, y, y', the components of $\mathbf{A}$ and the (extended) generators $\mathbf{X}_1$ and $\mathbf{X}_2$ are given by

$$\begin{aligned} &\mathbf{A} \sim (0, \quad 0, \quad 1), \\ &\mathbf{X}_1 \sim (\mathbf{X}_1\varphi, \ \mathbf{X}_1\psi, \ \mathbf{X}_1\chi), \\ &\mathbf{X}_2 \sim (\mathbf{X}_2\varphi, \ \mathbf{X}_2\psi, \ \mathbf{X}_2\chi). \end{aligned} \tag{7.36}$$

Since the $\mathbf{X}_N$ are symmetries, we have $[\mathbf{X}_N, \mathbf{A}] = \lambda_N\mathbf{A}$, and this implies that $\mathbf{X}_N\varphi$ and $\mathbf{X}_N\psi$ are independent of χ (they are again solutions and therefore functions only of φ and ψ). If we perform "coordinate" transformations $\phi = \phi(\varphi, \psi)$ and $\Psi = \Psi(\varphi, \psi)$ in the space of solutions (of first integrals to $y'' = \omega$), the components $\mathbf{X}_N\varphi$ and $\mathbf{X}_N\psi$ will change, and it makes sense to ask for normal forms of the (components of the extended) generators $\mathbf{X}_N$ in the space of solutions. This space is two-dimensional, like the space of variables x, y, and since the extended generators $\mathbf{X}_N$ have the same commutators as the nonextended $\mathbf{X}_N$, the classification of normal forms in the solution space is a simple repetition of what we have done in Section 7.1, and we shall end up with a complete analogue of Table 7.1 (except that there is no differential equation to be solved when we already have two first integrals).

We do not repeat this table here since for the following we shall need only some parts of those earlier results. We are mainly interested in the cases that are the analogues of Ia and IIa. These cases are characterized by the generator spanning a two-dimensional space; that is, they are transitive in (s, t) (or x, y)-space. In these cases a function u always exists such that $\mathbf{X}_1 u = 0$, $\mathbf{X}_2 u = 1$ ($u = t$ for Ia, $u = \ln t$ for IIa). To translate this into the notation of the solution space, we remark that because of (6.53) transitivity in solution space is guaranteed if the determinant Δ of the components of the three vectors $\mathbf{A}$, $\mathbf{X}_1$, and $\mathbf{X}_2$ does not vanish. This statement is independent of the coordinates; it is true in coordinates (φ, ψ, χ) as well as in coordinates (x, y, y'). If $\Delta \neq 0$, then – as shown in the preceding section – a solution φ (the analogue of the function u) with $\mathbf{X}_1\varphi = 0$, $\mathbf{X}_2\varphi = 1$ always exists.

So we can state: if two symmetries with (extended) generators $\mathbf{X}_1$ and $\mathbf{X}_2$ [satisfying (7.3) or (7.4)] are transitive in the space of solutions, that is, if

$$\Delta = \begin{vmatrix} 1 & y' & \omega \\ \xi_1 & \eta_1 & \eta_1' \\ \xi_2 & \eta_2 & \eta_2' \end{vmatrix} \neq 0, \tag{7.37}$$

then the system

$$\begin{aligned} \mathbf{A}\varphi &= \varphi_{,x} + y'\varphi_{,y} + \omega\varphi_{,y'} = 0, \\ \mathbf{X}_1\varphi &= \xi_1\varphi_{,x} + \eta_1\varphi_{,y} + \eta_1'\varphi_{,y'} = 0, \\ \mathbf{X}_2\varphi &= \xi_2\varphi_{,x} + \eta_2\varphi_{,y} + \eta_2'\varphi_{,y'} = 1, \end{aligned} \tag{7.38}$$

always has a solution $\varphi(x, y, y')$ that is unique (up to an additive constant).

The surprise the symmetry approach has in store is that the system (7.38) not only defines a first integral φ but also provides a simple way of obtaining it explicitly: since the determinant Δ of the system (7.38) is nonzero by assumption (of transitivity), the system can be solved with respect to $\varphi_{,x}$, $\varphi_{,y}$, and $\varphi_{,y'}$, and φ can be written in the form of a line integral:

$$\varphi(x, y, y') = \int (\varphi_{,x}\,\mathrm{d}x + \varphi_{,y}\,\mathrm{d}y + \varphi_{,y'}\,\mathrm{d}y') = \int \frac{\begin{vmatrix} \mathrm{d}x & \mathrm{d}y & \mathrm{d}y' \\ 1 & y' & \omega \\ \xi_1 & \eta_1 & \eta_1' \end{vmatrix}}{\Delta}. \tag{7.39}$$

We now have to hand one of the two first integrals, and what we still need is the second solution ψ of $\mathbf{A}f = 0$. If we introduce φ as a new variable instead of y' ($\mathbf{A}\varphi = 0$ ensures $\varphi_{,y'} \neq 0$ for nonconstant φ), the equation $\mathbf{A}\psi = 0$ reads

$$\mathbf{A}\psi = \left(\frac{\partial}{\partial x} + y'(x, y, \varphi)\frac{\partial}{\partial y}\right)\psi = 0, \tag{7.40}$$

and a symmetry it still admits is

$$\mathbf{X}_1 = \xi_1(x, y)\frac{\partial}{\partial x} + \eta_1(x, y)\frac{\partial}{\partial y}. \tag{7.41}$$

Again we encounter a pleasant surprise: equation (7.40) has exactly the form typical for a *first* order equation (5.1), φ only plays the role of an additional parameter, and since (7.40) admits the symmetry (7.41), we can use the results of Section 5.1 and give the solution ψ in the form of a line integral:

$$\psi(x, y, \varphi) = \int \frac{\mathrm{d}y - y'(x, y, \varphi)\,\mathrm{d}x}{\eta_1 - \xi_1 y'}. \tag{7.42}$$

We now have two first integrals φ and ψ, and from $\psi(x, y, \varphi_0) = \psi_0$ we can obtain $y = y(x, \varphi_0, \psi_0)$. So solving a second order differential equation with a G_2 of symmetries that acts transitively in the solution space amounts to performing two line integrals (7.39) and (7.42)!

If the two generators $\mathbf{X}_1$ and $\mathbf{X}_2$ commute ($[\mathbf{X}_1, \mathbf{X}_2] = 0$), we could have repeated the reasoning that led to (7.38) with φ and ψ and $\mathbf{X}_1$ and $\mathbf{X}_2$ interchanged. We would then have arrived at

$$\begin{aligned} \mathbf{A}\psi &= \psi_{,x} + y'\psi_{,y} + \omega\psi_{,y'} = 0, \\ \mathbf{X}_2\psi &= \xi_2\psi_{,x} + \eta_2\psi_{,y} + \eta'_2\psi_{,y'} = 0, \\ \mathbf{X}_1\psi &= \xi_1\psi_{,x} + \eta_1\psi_{,y} + \eta'_1\psi_{,y'} = 1, \end{aligned} \tag{7.43}$$

and have obtained

$$\psi(x, y, y') = \int (\psi_{,x}\,\mathrm{d}x + \psi_{,y}\,\mathrm{d}y + \psi_{,y'}\,\mathrm{d}y') = \int \frac{\begin{vmatrix} \mathrm{d}x & \mathrm{d}y & \mathrm{d}y' \\ 1 & y' & \omega \\ \xi_2 & \eta_2 & \eta'_2 \end{vmatrix}}{\Delta}, \tag{7.44}$$

which together with the line integral (7.39) represents the solution.

If the (extended) generators are *not* transitive in the solution space, that is, if $\Delta = 0$, then the integration procedure presented above breaks down, and we have ruefully to go back to the first integration strategy. When can $\Delta = 0$

happen? Consulting Table 7.1, we see, for example, that in case Ia we have

$$\Delta = \begin{vmatrix} 1 & s' & \hat{\omega} \\ 0 & 1 & 0 \\ 1 & 0 & 0 \end{vmatrix} = -\hat{\omega}; \tag{7.45}$$

so $\Delta = 0$ is possible only if $\hat{\omega} = 0$. In cases Ib and IIb, $\Delta = 0$ is impossible, but $\Delta = 0$ will happen again in case IIa for $\hat{\omega} = 0$. Note that the generators can be intransitive in the solution space only in a small subset of the cases where they are transitive in coordinate space (s, t). But note also that for $\hat{\omega} = 0$ an *eight*-parameter group of symmetries exists, among which there are always two with $\Delta \neq 0$ (and which form a group).

When the generators are intransitive in the solution space, a linear relation must hold between the operators $\mathbf{A}$, $\mathbf{X}_1$, and $\mathbf{X}_2$. As shown in Exercise 7.1, this relation can always be written as

$$\mathbf{X}_2 = \psi \mathbf{X}_1 + \nu \mathbf{A}, \tag{7.46}$$

with nonconstant ψ and ν. Taking the commutator with $\mathbf{A}$ of both sides of (7.46), we obtain

$$[\mathbf{A}, \mathbf{X}_2] = -\lambda_2 \mathbf{A} = (\mathbf{A}\psi)\mathbf{X}_1 - \psi\lambda_1 \mathbf{A} + (\mathbf{A}\nu)\mathbf{A}. \tag{7.47}$$

Since a linear relation between only *two* of the $\mathbf{A}, \mathbf{X}_1, \mathbf{X}_2$ is impossible (see Exercise 7.2), we infer from (7.47) that

$$\mathbf{A}\psi = 0. \tag{7.48}$$

Here $\psi(x, y, y')$ is a first integral, a solution to $\mathbf{A}f = 0$. So even when the integration procedure fails because $\Delta = 0$, one solution ψ can be found by simple algebraic manipulations – in analogy with the construction of $t(x, y)$ and $s(x, y)$ from (7.20) or (7.31), respectively.

7.4 Summary: Recipe for integration of second order differential equations admitting a group G_2

As discussed in the preceding sections, two different integration strategies can be used, and it is sometimes a matter of taste which to choose. Roughly speaking, in both approaches the same integrals have to be performed, but in the first strategy (using the normal forms of the generators in the space of variables) the task is divided into more and therefore smaller pieces, which are often easier to deal with.

In any case, one has to start with some classification work. Write the given differential equation $y'' = \omega(x, y, y')$ and its two known symmetries as

$$\mathbf{A} = \frac{\partial}{\partial x} + y'\frac{\partial}{\partial y} + \omega\frac{\partial}{\partial y'},$$

$$\mathbf{X}_1 = \xi_1\frac{\partial}{\partial x} + \eta_1\frac{\partial}{\partial y} + \eta_1'\frac{\partial}{\partial y'}, \tag{7.49}$$

$$\mathbf{X}_2 = \xi_2\frac{\partial}{\partial x} + \eta_2\frac{\partial}{\partial y} + \eta_2'\frac{\partial}{\partial y'}.$$

Then check for each of the following conditions which alternative holds:

$$[\mathbf{X}_1, \mathbf{X}_2] = \begin{cases} 0, \\ \mathbf{X}_1, \end{cases} \qquad \delta = \begin{vmatrix} \xi_1 & \eta_1 \\ \xi_2 & \eta_2 \end{vmatrix} \begin{cases} = 0, \\ \neq 0, \end{cases}$$

$$\Delta = \begin{vmatrix} 1 & y' & \omega \\ \xi_1 & \eta_1 & \eta_1' \\ \xi_2 & \eta_2 & \eta_2' \end{vmatrix} \begin{cases} = 0, \\ \neq 0. \end{cases} \tag{7.50}$$

If $\Delta \neq 0$, one can either calculate φ according to (7.39) and then ψ according to (7.42) or consult Table 7.1 to see which case applies and then go to the detailed prescription given for each case in Section 7.2. For commuting generators, one can also use (7.39) as above and (7.44) instead of (7.42) to obtain two first integrals φ and ψ.

For $[\mathbf{X}_1, \mathbf{X}_2] = \mathbf{X}_1$ and $\delta = 0$ (case IIb), the approach using (7.39) and (7.42) should always be chosen.

If $\Delta = 0$, which can happen only for $\delta \neq 0$, the way using the first strategy (cases Ia and IIa of Section 7.2) always has to be taken. Sometimes, the first integral ψ from (7.46) may also be useful. But one should check the list of symmetries: a G_8 is known to exist, which also contains a G_2 with $\Delta \neq 0$!

In any case, the task of integration has been reduced to some line integrals, elimination and transformations of variables, and quadratures: in this sense, the existence of a G_2 of symmetries gives an algorithmic way for solving a second order differential equation.

7.5 Examples

1. The differential equation

$$y'' = (x - y)y'^3 \tag{7.51}$$

admits (among others) the symmetries

$$\mathbf{X}_1 = \frac{\partial}{\partial x} + \frac{\partial}{\partial y}, \tag{7.52}$$

$$\mathbf{X}_2 = (x-y)\frac{\partial}{\partial x} + y'(y'-1)\frac{\partial}{\partial y'},$$

compare Section 4.3. We find that $\mathbf{X}_1$ and $\mathbf{X}_2$ commute and that δ and Δ do not vanish:

$$[\mathbf{X}_1, \mathbf{X}_2] = 0, \quad \delta = y - x, \quad \Delta = -y'^3(x-y)^2 - y'(1-y')^2. \quad (7.53)$$

If we apply the second integration strategy, that is, use (7.39) and (7.44), we obtain

$$\varphi = \int \frac{y'^3(x-y)\mathrm{d}x - y'^3(x-y)\mathrm{d}y - (1-y')\mathrm{d}y'}{y'^3(x-y)^2 + y'(1-y')^2}$$
$$= \frac{1}{2}\int\frac{\mathrm{d}\Delta}{\Delta} - \frac{3}{2}\int\frac{\mathrm{d}y'}{y'} = \tfrac{1}{2}\ln\left[(x-y)^2 + \frac{(1-y')^2}{y'^2}\right] \quad (7.54)$$

and

$$\psi = -\int \frac{y'^2(y'-1)\mathrm{d}x + [(x-y)^2 y'^3 - y'(y'-1)]\mathrm{d}y - y'(x-y)\mathrm{d}y'}{y'^3(x-y)^2 + y'(1-y')^2}$$
$$= -\int \mathrm{d}y - \int \frac{y'(y'-1)\mathrm{d}(x-y) - (x-y)\mathrm{d}y'}{y'^2(x-y)^2 + (1-y')^2}$$
$$= -y - \arctan\frac{y'(x-y)}{1-y'}. \quad (7.55)$$

From $\psi = \psi_0$ we get $\tan(y+\psi_0) = (y-x)y'/(1-y')$, and $\varphi = \varphi_0$ then leads to the implicit representation

$$x - y \pm e^{\varphi_0}\sin(y+\psi_0) = 0 \quad (7.56)$$

of the solution $y = y(x, \varphi_0, \psi_0)$.

To compare the two approaches, we apply the first integration strategy to the same example. Because of (7.53), we have case Ia of Table 7.1 and Section 7.2. We can obtain the functions $t(x,y)$ and $s(x,y)$ from (7.11) and (7.12), that is, from

$$t(x,y) = \int \frac{-\mathrm{d}x + \mathrm{d}y}{y-x} = \ln(x-y), \quad (7.57)$$

$$s(x,y) = \int \frac{-(x-y)\,\mathrm{d}y}{y-x} = y. \quad (7.58)$$

To write the differential equation in terms of s and t, we have to compute y' from (7.17), which yields

$$y' = \frac{s'}{e^t + s'}, \qquad s' = \frac{y'(x-y)}{1-y'}, \tag{7.59}$$

and then

$$s''(t) = s' + s'^3. \tag{7.60}$$

A first integral of this equation is

$$s'^2 = e^{2(t-\varphi_0)}/(1 - e^{2(t-\varphi_0)}), \tag{7.61}$$

the analogue of (7.54). If again integrated, we obtain

$$s + \psi_0 = \pm \arcsin e^{t-\varphi_0}, \tag{7.62}$$

which together with (7.57) and (7.58) yields (7.56), as it should.

This last approach avoids the somewhat tedious line integrals, but the number of steps to be gone through is clearly larger. Both approaches fail to detect (and to use) the fact that the differential equation $y'' = (x-y)y'^3$ can be transformed into $y'' = -y + x$ by a simple change of variables $x \leftrightarrow y$ but nevertheless arrive at the correct result.

2. The differential equation

$$y'' = x^{-5}y^2 \tag{7.63}$$

admits the two symmetries

$$\begin{aligned} \mathbf{X}_1 &= -x^2\frac{\partial}{\partial x} - xy\frac{\partial}{\partial y} + (xy' - y)\frac{\partial}{\partial y'}, \\ \mathbf{X}_2 &= -x\frac{\partial}{\partial x} - 3y\frac{\partial}{\partial y} - 2y'\frac{\partial}{\partial y'}. \end{aligned} \tag{7.64}$$

The generators do not commute, and δ and Δ do not vanish:

$$[\mathbf{X}_1, \mathbf{X}_2] = \mathbf{X}_1, \quad \delta = 2x^2y, \quad \Delta = -3y^2 - 3x^2y'^2 + 2x^{-3}y^3 + 6xyy'. \tag{7.65}$$

We decide to use equations (7.39) and (7.42) for the integration procedure. From (7.39) we have

$$\begin{aligned}\varphi(x, y, y') &\\ &= -\int\frac{[y'(y-xy')-x^{-4}y^3]\,\mathrm{d}x+[xy'-y+x^{-3}y^2]\,\mathrm{d}y+(xy-x^2y')\,\mathrm{d}y'}{2x^{-3}y^3-3x^2y'^2-3y^2+6xyy'}\\ &= -\tfrac{1}{6}\ln\Delta = -\tfrac{1}{6}\ln[2x^{-3}y^3-3x^2y'^2-3y^2+6xyy'].\end{aligned} \tag{7.66}$$

This equation is to be resolved for y'. It yields (with $\varphi=\varphi_0$)

$$y'(x, y, \varphi_0) = \frac{y}{x} \pm \left[\frac{2}{3}\frac{y^3}{x^5} - \frac{e^{-6\varphi_0}}{x^2}\right]^{1/2}. \tag{7.67}$$

The line integral (7.42) for $\psi(x, y, \varphi_0)$ then reads

$$\begin{aligned}\psi(x, y, \varphi_0) &= \int\frac{\mathrm{d}y - y'\,\mathrm{d}x}{-xy-x^2y'}\\ &= \int\frac{\mathrm{d}x}{x^2} \pm \int\frac{\mathrm{d}y-(y/x)\mathrm{d}x}{x^2[\frac{2}{3}(y^3/x^5)-e^{-6\varphi_0}/x^2]^{1/2}}.\end{aligned} \tag{7.68}$$

The form of the second integrand indicates that we should introduce $u=y/x$ as a new variable. Doing this, we obtain

$$\psi(x, y, \varphi_0) = -\frac{1}{x} \pm \int\frac{\mathrm{d}u}{[\frac{2}{3}u^3 - e^{-6}\varphi_0]^{1/2}}, \qquad u \equiv \frac{y}{x}, \tag{7.69}$$

and from $\psi=\psi_0$ we could get $y(x, \varphi_0, \psi_0)$ in terms of elliptic integrals (the constant $e^{-6\varphi_0}$ need not in reality be positive; we could have assigned the name φ_0 to that constant from the beginning).

One remark should be added concerning the interconnection of the two main integration strategies: the variables x^{-1} and y/x that quite naturally occur in (7.69) are closely related to the normal form (IIa) of the operators since $s=x^{-1}$ and $t^2=y/x$ would follow from (7.26) and (7.28).

3. The differential equation

$$y'' = 2y' - y + (y'-y)/x \tag{7.70}$$

admits (among others) the two commuting symmetries

$$\begin{aligned}\mathbf{X}_1 &= e^x\frac{\partial}{\partial y} + e^x\frac{\partial}{\partial y'},\\ \mathbf{X}_2 &= \frac{1}{2x}\frac{\partial}{\partial x} + \frac{y}{2x}\frac{\partial}{\partial y} + \left(\frac{y'-y}{x} - y'\right)\frac{1}{2x}\frac{\partial}{\partial y'},\end{aligned} \tag{7.71}$$

for which we find

$$[\mathbf{X}_1, \mathbf{X}_2] = 0, \qquad \delta = -\frac{e^x}{2x}, \qquad \Delta = 0. \tag{7.72}$$

Because $\Delta = 0$, we know that the differential equation can be transformed into $s'' = 0$ and that from the linear relation (7.46) between $\mathbf{A}$, $\mathbf{X}_1$, and $\mathbf{X}_2$ a first integral ψ can be read off. We find

$$\psi(x, y, y') = \frac{e^{-x}}{2x}(y - y') = \psi_0. \tag{7.73}$$

The line integrals (7.13) and (7.14) yield

$$t(x, y) = x^2, \qquad s(x, y) = ye^{-x}, \tag{7.74}$$

and because $s'' = 0$, the solution is $s = at + b$, that is,

$$y = e^x(ax^2 + b). \tag{7.75}$$

7.6 Exercises

1 Show that the function s defined by (7.32) satisfies $\mathbf{X}_1 s = 1$.

2 Show that if a linear relation $\mu_1 \mathbf{X}_1 + \mu_2 \mathbf{X}_2 + \nu \mathbf{A} = 0$ holds, then μ_1, μ_2, and ν cannot be constants.

3 Show that the differential equation $y'' = x^{-15/7} y^2$, with the symmetries given in Table 4.1, has the solution

$$\psi_0 = \tfrac{12}{49} x^{1/7} + \frac{12}{7^3} \int \frac{-x^{-1/7}\, du + x^{6/7}\left(\frac{2}{7}u + \frac{12}{7^3}\right) dx}{\left[\frac{4}{7^3} e^{-6\varphi_0} x^{6/7} + \frac{2}{3}u^3 + \frac{12}{49}u^2 + \frac{72}{7^4}u + \frac{12^2}{7^6}\right]^{1/2}},$$

$$u \equiv yx^{-1/7}.$$

4 Show that if for two generators $\mathbf{X}_1$, $\mathbf{X}_2$ (which form a Lie algebra) $\delta_1 = \xi_1 \eta_2 - \xi_2 \eta_1 = 0$ holds, then the differential equation $y'' = \omega(x, y, y')$ can be transformed into $d^2\hat{s}/dt^2 = 0$; it admits eight symmetries.

5 Solve $y'' = y + x^2$ by using its symmetries $\mathbf{X}_1 = e^x\, \partial/\partial y$, $\mathbf{X}_2 = e^{-x}\, \partial/\partial y$.

8

Second order differential equations admitting more than two Lie point symmetries

8.1 The problem: groups that do not contain a G_2

A second order differential equation can admit up to eight Lie point symmetries: see Section 4.3 (if there really are eight symmetries, then the differential equation can be transformed into $y''=0$). The existence of a G_2 of symmetries is already sufficient to make an integration procedure via line integrations possible. The more symmetries, the easier the task; so with more than two symmetries, we should expect even less trouble, should we not?

Generally speaking, we face the following problem. The group G_r, $2 < r \leqslant 8$, of symmetries has a certain structure expressed in the structure constants. Any integration strategy we could invent depends on this structure, and to cover all possible cases, we should make a list of all different groups that admit a realization in the x–y plane and invent a strategy for each of them. That is possible, but not really practicable: if the G_r contains a subgroup G_2, then we can forget the rest, take those two symmetries, and integrate straightforwardly. Only if the G_r does not contain a subgroup G_2 do we have to invent something new.

Which G_r do not contain a G_2? Lie already gave a complete list of all possible Lie groups acting in the x–y plane, and from this list we can infer that there exists only one group that does not contain a G_2, and this is the three-parameter group G_3 with commutators

$$[\mathbf{X}_1, \mathbf{X}_2] = \mathbf{X}_3, \qquad [\mathbf{X}_2, \mathbf{X}_3] = \mathbf{X}_1, \qquad [\mathbf{X}_3, \mathbf{X}_1] = \mathbf{X}_2; \tag{8.1}$$

that is, the group isomorphic to the group of rotations of a three-dimensional space [compare (6.37) where this group was discussed as an example of a group that is not solvable].

So if we find a group G_r, $r > 2$, of symmetries of a second order differential equation, either we shall also find a subgroup G_2 (and then we can use the results of Chapter 7) or we have to deal with the group with structure constants given by (8.1), which is called G_3IX in the Bianchi–Lie classification of groups G_3. To find a (the) subgroup G_2 or to show its nonexistence may be lengthy but is a straightforward task; compare Exercise 8.4.

8.2 How to solve differential equations that admit a G_3IX

The integration strategy for second order differential equations that admit a G_3IX with commutators (8.1) is almost trivial. Since in the space of *three* variables (x, y, y') the *four* operators $\mathbf{A}$, $\mathbf{X}_N$ are always linearly dependent, a relation

$$\mu_1 \mathbf{X}_1 + \mu_2 \mathbf{X}_2 + \mu_3 \mathbf{X}_3 + \nu \mathbf{A} = 0 \tag{8.2}$$

must hold. The recipe now is: write this equation as

$$\mathbf{X}_1 = \varphi(x, y, y')\mathbf{X}_2 + \psi(x, y, y')\mathbf{X}_3 + v(x, y, y')\mathbf{A}. \tag{8.3}$$

Then the functions φ and ψ are two independent first integrals satisfying

$$\mathbf{A}\varphi = 0, \qquad \mathbf{A}\psi = 0, \tag{8.4}$$

and from $\varphi = \varphi_0$ and $\psi = \psi_0$ one can obtain the solution $y(x, \varphi_0, \psi_0)$. No integration is necessary at all; only some algebra has to be done! [Of course, one could alternatively solve (8.2) for $\mathbf{X}_2$ or $\mathbf{X}_3$.]

To prove this assertion, we shall show that none of the μ_i in equation (8.2) can be zero, or constant, that φ and ψ satisfy (8.4), and that φ and ψ are functionally independent.

To show that none of the μ_i can vanish, we assume that a relation

$$\mu_2 \mathbf{X}_2 + \mu_3 \mathbf{X}_3 + \nu \mathbf{A} = 0 \tag{8.5}$$

holds. If we introduce variables $\phi(x, y, y')$, $\Psi(x, y, y')$, and $\chi(x, y, y')$ with

$$\mathbf{A}\phi = \mathbf{A}\Psi = 0, \qquad \mathbf{A}\chi = 1 \tag{8.6}$$

(compare the beginning of Section 7.3), then we have

$$\mathbf{X}_2 = (\mathbf{X}_2\phi)\frac{\partial}{\partial \phi} + (\mathbf{X}_2\Psi)\frac{\partial}{\partial \Psi} + (\mathbf{X}_2\chi)\frac{\partial}{\partial \chi},$$

$$\mathbf{X}_3 = (\mathbf{X}_3\phi)\frac{\partial}{\partial\phi} + (\mathbf{X}_3\Psi)\frac{\partial}{\partial\Psi} + (\mathbf{X}_3\chi)\frac{\partial}{\partial\chi}, \tag{8.7}$$

$$\mathbf{A} = \qquad\qquad\qquad\qquad\qquad\qquad \frac{\partial}{\partial\chi},$$

and we know that because of $[\mathbf{X}_i, \mathbf{A}] = \lambda_i\mathbf{A}$, none of the $\mathbf{X}_i\phi$ or $\mathbf{X}_i\Psi$ depends on χ. That is, a two-dimensional realization of the Lie algebra in the space of solutions exists. If we now transform, say, $\mathbf{X}_2$ into its normal form in this space, then we have, for example, $\mathbf{X}_2\phi = 1, \mathbf{X}_2\Psi = 0$ (with some new ϕ, Ψ of course); $\mathbf{X}_2\phi = 0 = \mathbf{X}_2\Psi$ would imply that $\mathbf{X}_2$ is proportional to $\mathbf{A}$ (compare Section 3.2), which is impossible for a second order differential equation. Since the determinant of the system (8.7) must be zero, $\mathbf{X}_3\Psi$ will also vanish, and because of the commutator relation $[\mathbf{X}_2, \mathbf{X}_3] = \mathbf{X}_1\Psi$ is also zero. So we have $\mathbf{X}_i\Psi = 0$; a one-dimensional realization of the group in the ϕ-space must exist – which contradicts the fact that the group under consideration does not possess any one-dimensional realization, as shown in Section 6.4. That means that no linear relation (8.5) between $\mathbf{A}$ and two of the $\mathbf{X}_i$ exists. Or stated differently: if a linear relation of the form (8.5) exists, then all the coefficients μ_2, μ_3, and ν must vanish.

We now take the relation (8.3) and form the commutator of both sides with $\mathbf{A}$. The result is

$$-\lambda_1\mathbf{A} = (\mathbf{A}\varphi)\mathbf{X}_2 - \lambda_2\mathbf{A} + (\mathbf{A}\psi)\mathbf{X}_3 - \lambda_3\mathbf{A} + (\mathbf{A}\nu)\mathbf{A}. \tag{8.8}$$

This is a linear relation between $\mathbf{X}_2$, $\mathbf{X}_3$, and $\mathbf{A}$ so the coefficients must vanish, which yields

$$\mathbf{A}\varphi = 0 = \mathbf{A}\psi. \tag{8.9}$$

Taking the commutators of both sides of (8.3) with $\mathbf{X}_1$, $\mathbf{X}_2$, and $\mathbf{X}_3$ and substituting $\varphi\mathbf{X}_2 + \psi\mathbf{X}_3 + \nu\mathbf{A}$ for any new $\mathbf{X}_1$, the same reasoning leads to

$$\begin{aligned} \mathbf{X}_1\varphi &= \psi, & \mathbf{X}_1\psi &= -\varphi, \\ \mathbf{X}_2\varphi &= -\varphi\psi, & \mathbf{X}_2\psi &= -1-\psi^2, \\ \mathbf{X}_3\varphi &= 1+\varphi^2, & \mathbf{X}_3\psi &= \varphi\psi. \end{aligned} \tag{8.10}$$

If φ or ψ or both were constant, these equations could not be satisfied. If φ were a function of ψ, the system (8.10) would give a one-dimensional realization of the group, which does not exist [moreover, (8.10) would lead to a contradiction]. So we know that φ and ψ are independent functions, that is, because of (8.9), two independent first integrals of the given differential equation.

By the way, for other groups with more than two generators it always pays to write down linear relation(s) of the type (8.3). The coefficients φ and ψ will always satisfy $\mathbf{A}\varphi = 0 = \mathbf{A}\psi$ (if no linear relation between $\mathbf{X}_2, \mathbf{X}_3$, and $\mathbf{A}$ exists), but they need not be nonconstant or independent functions.

8.3 Example

The differential equation

$$y'' = 2y'^2 \cot y + \sin y \cos y \tag{8.11}$$

of the geodesics on the sphere admits the generators

$$\begin{aligned}
\mathbf{X}_1 &= \frac{\partial}{\partial x},\\
\mathbf{X}_2 &= \cot y \cos x \frac{\partial}{\partial x} + \sin x \frac{\partial}{\partial y}\\
&\quad + \left(\cos x + y' \cot y \sin x + y'^2 \frac{\cos x}{\sin^2 y}\right)\frac{\partial}{\partial y'},\\
\mathbf{X}_3 &= -\cot y \sin x \frac{\partial}{\partial x} + \cos x \frac{\partial}{\partial y}\\
&\quad + \left(-\sin x + y' \cot y \cos x - y'^2 \frac{\sin x}{\sin^2 y}\right)\frac{\partial}{\partial y'}
\end{aligned} \tag{8.12}$$

of the three-dimensional rotation group. The coefficients φ and ψ appearing in the linear relation $\mathbf{X}_1 = \varphi\mathbf{X}_2 + \psi\mathbf{X}_3 + \nu\mathbf{A}$ between these generators and

$$\mathbf{A} = \frac{\partial}{\partial x} + y'\frac{\partial}{\partial y} + (2y'^2 \cot y + \sin y \cos y)\frac{\partial}{\partial y'} \tag{8.13}$$

can be computed by using some matrices and determinants routine. The result is

$$\begin{aligned}
\varphi &= -y'\frac{\sin x}{\sin^2 y} - \cos x \cot y,\\
\psi &= -y'\frac{\cos x}{\sin^2 y} + \sin x \cot y.
\end{aligned} \tag{8.14}$$

If we eliminate y' from $\varphi = \varphi_0$, $\psi = \psi_0$, we obtain the solution to (8.11) as

$$\cot y = \psi_0 \sin x - \varphi_0 \cos x, \tag{8.15}$$

8.4 Exercises

1 Show that the Lie algebra with commutators

$$[\mathbf{X}_1, \mathbf{X}_2] = -\mathbf{X}_3, \qquad [\mathbf{X}_1, \mathbf{X}_3] = -\mathbf{X}_2, \qquad [\mathbf{X}_2, \mathbf{X}_3] = \mathbf{X}_1$$

has a two-dimensional subalgebra.

2 The differential equation $3yy'' = 5y'^2$ admits the three symmetries $\mathbf{X}_1 = \partial/\partial x$, $\mathbf{X}_2 = x\,\partial/\partial x$, $\mathbf{X}_3 = y\,\partial/\partial y$. Use them to integrate it. Does (8.3) provide first integrals?

3 Show that the group G_3IX with commutators (8.1) is a subgroup of the projective group with generators (4.26). (*Hint*: use one of the two realizations of G_3IX given above.)

4 The differential equation $y'' = 2(xy' - y)/x^2$ admits the three symmetries $\mathbf{X}_1 = y\,\partial/\partial x - x\,\partial/\partial y$, $\mathbf{X}_2 = -(y/x)\,\partial/\partial x + (x^2 - y^2/x^2)\,\partial/\partial y$, $\mathbf{X}_3 = (1 + x^2)\,\partial/\partial x + (2xy + y/x)\,\partial/\partial y$. Does the group possess a two-dimensional subgroup with generators $\mathbf{Y}_\alpha = a_\alpha^i \mathbf{X}_i$, $\alpha = 1, 2$? Integrate the differential equation by using the symmetries!

9

Higher order differential equations admitting more than one Lie point symmetry

9.1 The problem: some general remarks

As already stated in Section 6.5, it would be an infinite program to cover all possible combinations of nth order differential equations ($n = 3, 4, \ldots$) and groups $G_r (r = 1, \ldots, n+4)$ and to find the optimal integration strategy for each of them. So we shall concentrate on a few topics, which together may give an illustration of what can be done. Before we start to do this in the following sections, we make some remarks concerning the main ideas on which the approaches will be based.

The experience of differential equations with one symmetry, which could either be solved (if they were of first order) or could be reduced in order by one, and of the second order differential equations with two symmetries, which could be solved, may lead us to the following conjecture:

nth order differential equation $+ r$ symmetries ($r \leqslant n$)

$= (n - r)$th order differential equation + something easy to perform.

Is this conjecture true? In full generality, no. But it is exactly the goal of all integration strategies to make it as true as possible, which means to make the "something" really simple and easy and the order of the differential equation to be solved as low as possible. Whether this is feasible or not very much depends on the structure of the group G_r of symmetries. To get a deeper insight, we need much more knowledge about Lie algebras than we have collected in Chapter 6. So we shall discuss only some simple strategies and refer readers to the literature, where they may look under the headings simple groups, adjoint representation, derived group, invariant subgroup, and characteristic equation.

Of course, the formulation of the conjecture is an invitation to proceed by induction, that is, to start from an nth order differential equation with r symmetries and try to reduce it to an $(n-1)$th order differential equation with $r-1$ symmetries. As for the second order differential equations, we shall consider two approaches, namely, to use normal forms of the operators in coordinate space (Section 9.2) or in the space of first integrals (Section 9.3). A different way of looking at the problem will be discussed in Section 9.4.

9.2 First integration strategy: normal forms of generators in the space(s) of variables

Suppose we have an nth order differential equation

$$y^{(n)} = \omega(x, y, y', \ldots, y^{(n-1)}), \tag{9.1}$$

which admits r symmetries $\mathbf{X}_N$. We take one of the generators, say, $\mathbf{X}_1$, and transform it to its normal form $\mathbf{X}_1 = \partial/\partial s$. We then know from Section 5.2 that this amounts to having transformed the differential equation (9.1) into

$$s^{(n)} = \tilde{\omega}(t, s', s'', \ldots, s^{(n-1)}), \tag{9.2}$$

which in fact is a differential equation of order $n-1$ for $s'(t)$ [and that to perform this transformation, we must solve a first order differential equation – which was promised to be simple – and deal with the two quadratures (5.16) and (5.22) or (5.23)].

Can we use one of the remaining symmetries $\mathbf{X}_2, \ldots, \mathbf{X}_r$ to repeat this procedure for the differential equation (9.2)? Did (9.2) really inherit these symmetries from (9.1)?

To answer this question, we analyse a bit more carefully what we have done. The differential equation (9.1) and its symmetries are (in coordinates t, s) represented by the operators

$$\begin{aligned}
\mathbf{A} &= \frac{\partial}{\partial t} + s'\frac{\partial}{\partial s} + s''\frac{\partial}{\partial s'} + \cdots + \tilde{\omega}\frac{\partial}{\partial s^{(n-1)}}, \\
\mathbf{X}_1 &= \frac{\partial}{\partial s}, \\
\mathbf{X}_a &= \xi_a\frac{\partial}{\partial t} + \eta_a\frac{\partial}{\partial s} + \eta'_a\frac{\partial}{\partial s'} + \cdots + \eta_a^{(n-1)}\frac{\partial}{\partial s^{(n-1)}}, \qquad a = 2, \ldots, r.
\end{aligned} \tag{9.3}$$

Solving the original equation means finding n first integrals $\varphi^\alpha(t, s, s', \ldots, s^{(n-1)})$ satisfying $\mathbf{A}\varphi^\alpha = 0$. Introducing s' as the dependent variable and then solving (9.2) means finding $n-1$ first integrals $\psi^\alpha(t, s', \ldots, s^{(n-1)})$ of $\mathbf{A}f = 0$ *that do not depend on s*. So in dealing with the reduced problem it is adequate to cancel all terms in the operators $\mathbf{A}$, $\mathbf{X}_a$ that contain $\partial/\partial s$ and instead of (9.3) to use

$$\hat{\mathbf{A}} = \frac{\partial}{\partial t} + s''\frac{\partial}{\partial s'} + \cdots + \tilde{\omega}(t, s', s'', \ldots)\frac{\partial}{\partial s^{(n-1)}},$$
$$\mathbf{Y}_a = \xi_a\frac{\partial}{\partial t} + \eta'_a\frac{\partial}{\partial s'} + \cdots + \eta_a^{(n-1)}\frac{\partial}{\partial s^{(n-1)}}. \tag{9.4}$$

The question now is whether the $\mathbf{Y}_a$ are symmetries of $\hat{\mathbf{A}}f = 0$, that is, whether because of

$$[\mathbf{X}_a, \mathbf{A}] = \lambda_a\mathbf{A} = -(\mathbf{A}\xi_a)\mathbf{A}, \tag{9.5}$$

$$[\mathbf{Y}_a, \hat{\mathbf{A}}] = \mu_a\hat{\mathbf{A}} = -(\hat{\mathbf{A}}\xi_a)\hat{\mathbf{A}} \tag{9.6}$$

also holds. To check this, we substitute $\mathbf{Y}_a + \eta_a\mathbf{X}_1$ for $\mathbf{X}_a$ and $\hat{\mathbf{A}} + s'\mathbf{X}_1$ for $\mathbf{A}$ in equation (9.5). We obtain

$$[\mathbf{Y}_a + \eta_a\mathbf{X}_1, s'\mathbf{X}_1 + \hat{\mathbf{A}}] = \eta'_a\mathbf{X}_1 - s'[\mathbf{X}_1, \mathbf{X}_a] + [\mathbf{Y}_a, \hat{\mathbf{A}}] - (\hat{\mathbf{A}}\eta_a)\mathbf{X}_1$$
$$= \lambda_a(\hat{\mathbf{A}} + s'\mathbf{X}_1). \tag{9.7}$$

To yield the desired form (9.6) for $[\mathbf{Y}_a, \hat{\mathbf{A}}]$, the commutator $[\mathbf{X}_1, \mathbf{X}_a]$ must be a function linear in $\mathbf{X}_1$ and $\hat{\mathbf{A}}$,

$$[\mathbf{X}_1, \mathbf{X}_a] = \tau_a\mathbf{X}_1 + \sigma_a\hat{\mathbf{A}}, \tag{9.8}$$

with some functions σ_a and τ_a. If we insert (9.6) and (9.8) into (9.7) and equate to zero the coefficients of $\mathbf{X}_1$ and $\hat{\mathbf{A}}$, we get

$$\eta'_a - s'\tau_a - \hat{\mathbf{A}}\eta_a = \lambda_a s' = -(\mathbf{A}\xi_a)s',$$
$$-s'\sigma_a + \mu_a = \lambda_a. \tag{9.9}$$

Using the definition $\eta'_a = (\mathbf{A}\eta_a) - s'(\mathbf{A}\xi_a)$ of η'_a, we obtain $\tau_a = \partial\eta_a/\partial s$, and because of the structure (9.3) of $\mathbf{X}_a$ and $\mathbf{A}$, this is consistent with (9.8) only if $\sigma_a = 0$. Then $[\mathbf{X}_1, \mathbf{X}_a] = \tau_a\mathbf{X}_1$ yields

$$\tau_a = \frac{\partial\eta_a}{\partial s}, \qquad \frac{\partial\xi_a}{\partial s} = \frac{\partial\eta'_a}{\partial s} = \cdots = \frac{\partial\eta_a^{(n-1)}}{\partial s} = 0, \tag{9.10}$$

and because of the definition of η'_a, this is possible only if $\tau_a = \text{const}$. The result of these computations is: the $\mathbf{Y}_a$ are symmetries of the reduced differential equation $\hat{\mathbf{A}}f = 0$ if and only if

$$[\mathbf{X}_1, \mathbf{X}_a] = \text{const.}\, \mathbf{X}_1 \tag{9.11}$$

holds, that is, all structure constants C^b_{1a} must vanish (for a, $b \neq 1$).

So if we want to follow this first integration strategy for a given algebra of generators, we should at the first step choose a generator $\mathbf{X}_1$ (as a linear combination of the given basis) for which we find as many generators $\mathbf{X}_a$ satisfying (9.11) as possible. There may be none at all! (Not every algebra has a subalgebra that contains an invariant one-dimensional algebra.) If we have chosen an $\mathbf{X}_1$, then we go from the $\mathbf{X}_a$ that obey (9.11) to the $\mathbf{Y}_a$ (from the group to the factor group), choose a $\mathbf{Y}_a$, and try to do everything again. At each step we have to introduce normal coordinates (to solve a simple first order differential equation), and then we can reduce the order of the given differential equation by one.

9.3 Second integration strategy: normal forms of generators in the space of first integrals. Lie's theorem

To transform generators to a normal form in the space of solutions (first integrals) anticipates that we have these solutions to hand or can provide the means to find them. Since we shall be able to find them only (if at all!) if we have enough symmetries, we assume from the beginning that the number of symmetries r at least equals the order n of the differential equation. In fact we want to start with $r = n$.

What we want to establish is an iterative procedure. In generalizing the method of Section 7.3, we begin with the following question: is it possible to find a first integral φ (a solution of $\mathbf{A}f = 0$) such that one of the generators, say, $\mathbf{X}_1$, can be written as $\partial/\partial\varphi$ and all other generators leave that first integral invariant? That is, does a (nonconstant) solution φ to the set of equations

$$\mathbf{X}_1\varphi = \left(\xi_1\frac{\partial}{\partial x} + \eta_1\frac{\partial}{\partial y} + \cdots + \eta_1^{(n-1)}\frac{\partial}{\partial y^{(n-1)}}\right)\varphi = 1 \tag{9.12}$$

and

$$\mathbf{X}_a\varphi = \left(\xi_a\frac{\partial}{\partial x} + \eta_a\frac{\partial}{\partial y} + \cdots + \eta_a^{(n-1)}\frac{\partial}{\partial y^{(n-1)}}\right)\varphi = 0,$$

$$a = 2, \dots, n, \tag{9.13}$$

$$\mathbf{A}\varphi = \left(\frac{\partial}{\partial x} + y'\frac{\partial}{\partial y} + \cdots + \omega(x, y, \dots, y^{(n-1)})\frac{\partial}{\partial y^{(n-1)}}\right)\varphi = 0$$

exist?

The system (9.13) is a system of n homogeneous linear partial differential equations in $n+1$ variables $(x, y, y', \ldots, y^{(n-1)})$. It will have a solution if its integrability conditions are fulfilled. These integrability conditions are that all commutators between $\{\mathbf{A}, \mathbf{X}_a\}$ are linear combinations (not necessarily with constant coefficients) of these same operators. Since the $\mathbf{X}_a$ are symmetries, we have $[\mathbf{X}_a, \mathbf{A}] = \lambda\mathbf{A}$; no extra conditions originate. Because the $\mathbf{X}_a$ are (some of the) generators of a group, we have

$$[\mathbf{X}_a, \mathbf{X}_b] = C^p_{ab}\mathbf{X}_p = C^d_{ab}\mathbf{X}_d + C^1_{ab}\mathbf{X}_1, \qquad a, b, d = 2, \ldots, n, \tag{9.14}$$

and the integrability conditions of (9.13) amount to the condition that the right side of (9.14) is a linear combination of the $\mathbf{X}_a$ and $\mathbf{A}$. If we assume (and we do so from now on) that *no* relation such as

$$\mathbf{X}_1 = \rho^a\mathbf{X}_a + \nu\mathbf{A} \tag{9.15}$$

holds, then the integrability conditions yield

$$C^1_{ab} = 0. \tag{9.16}$$

[By the way, if (9.15) were true, then the $\mathbf{X}_N$ would not be transitive in the space of first integrals.]

To ensure that $\mathbf{X}_1\varphi = 1$ does not contradict the system (9.13), we again have to satisfy integrability conditions. These are $[\mathbf{X}_1, \mathbf{A}]\varphi = 0$, which is true since $\mathbf{X}_1$ is a symmetry, and

$$[\mathbf{X}_1, \mathbf{X}_a]\varphi = (C^b_{1a}\mathbf{X}_b + C^1_{1a}\mathbf{X}_1)\varphi = 0. \tag{9.17}$$

Because $\mathbf{X}_1\varphi = 1$, this latter condition can hold only if $C^1_{1a} = 0$. Together with (9.16), this yields

$$[\mathbf{X}_1, \mathbf{X}_a] = C^b_{1a}\mathbf{X}_b, \qquad [\mathbf{X}_a, \mathbf{X}_b] = C^d_{ab}\mathbf{X}_d, \qquad C^1_{1a} = 0 = C^1_{ab}, \qquad d, a, b = 2, \ldots, n \tag{9.18}$$

(the $\mathbf{X}_a$ generate an invariant subgroup).

So the above question can be answered as follows: if the group is transitive in the space of first integrals [no relation (9.15) holds], then a solution φ to the system (9.12) and (9.13) exists if and only if (9.18) holds. But then this solution can be written in terms of a line integral: by assumption, $\mathbf{A}$ and the $\mathbf{X}_N$ are linearly independent, the determinant of the system does not vanish, and the derivatives of φ can be computed, and we obtain

$$\varphi(x, y, \ldots, y^{(n-1)}) = \int \frac{\begin{vmatrix} dx & dy & dy' & \cdots & dy^{(n-1)} \\ \xi_2 & \eta_2 & \eta_2' & \cdots & \eta_2^{(n-1)} \\ \vdots & \vdots & \vdots & & \vdots \\ \xi_n & \eta_n & \eta_n' & \cdots & \eta_n^{(n-1)} \\ 1 & y' & y'' & \cdots & \omega \end{vmatrix}}{\begin{vmatrix} \xi_1 & \eta_1 & \eta_1' & \cdots & \eta_1^{(n-1)} \\ \xi_2 & \eta_2 & \eta_2' & \cdots & \eta_2^{(n-1)} \\ \vdots & \vdots & \vdots & & \vdots \\ \xi_n & \eta_n & \eta_n' & \cdots & \eta_n^{(n-1)} \\ 1 & y' & y'' & \cdots & \omega \end{vmatrix}}, \tag{9.19}$$

the generalization of the line integral (7.39) used in integrating second order differential equations.

We can now use φ instead of $y^{(n-1)}$ as a new variable, and in the variables $\{\varphi, x, y, \ldots, y^{(n-2)}\}$ the operators $\mathbf{A}$, $\mathbf{X}_N$ are given by

$$\mathbf{X}_1 = \frac{\partial}{\partial \varphi} \tag{9.20}$$

and

$$\begin{aligned} \mathbf{X}_a &= \xi_a \frac{\partial}{\partial x} + \eta_a \frac{\partial}{\partial y} + \cdots + \eta_a^{(n-2)} \frac{\partial}{\partial y^{(n-2)}}, \\ \mathbf{A} &= \frac{\partial}{\partial x} + y' \frac{\partial}{\partial y} + \cdots + y^{(n-1)}(\varphi; x, y, \ldots, y^{(n-2)}) \frac{\partial}{\partial y^{(n-2)}}. \end{aligned} \tag{9.21}$$

In the operators $\mathbf{X}_a$, $\mathbf{A}$ no $\partial/\partial\varphi$ terms appear. The system (9.21) is exactly what we wanted to achieve: it corresponds to a differential equation $y^{(n-1)} = y^{(n-1)}(x, y, y', \ldots, y^{(n-2)})$ of order $n-1$, admitting the $n-1$ symmetries $\mathbf{X}_a$ (and depending on a – constant – parameter φ).

We can now again start to choose a generator, say, $\mathbf{X}_2$, and try to reduce the problem to an equation of order $n-2$ and then try to play the game again and again until we have found n first integrals and thus the complete solution. At each step we have to find a generator that satisfies the equivalent of equation (9.18); at the second step, for example,

$$[\mathbf{X}_2, \mathbf{X}_a] = C_{2a}^b \mathbf{X}_b, \qquad [\mathbf{X}_a, \mathbf{X}_b] = C_{ab}^d \mathbf{X}_d, \qquad d, a, b = 3, \ldots, n. \tag{9.22}$$

Is that possible? Do those generators exist? Is there a way to decide whether a given algebra permits us to find the necessary special generators at each step? The answer to the last question is yes, it is very simple to decide whether n such generators exist.

To prove this assertion, we take a closer look at the structure of the condition (9.18). If we write it as

$$[\mathbf{X}_N, \mathbf{X}_M] = C^P_{NM}\mathbf{X}_P, \qquad C^1_{NM} = 0, \tag{9.23}$$

we see that it demands that $\mathbf{X}_1$ does not appear at the right side of the commutators. But $\mathbf{X}_1$ was just a name for an arbitrary generator of the group, and so the condition is that the right side does not span the full algebra; at least one generator is missing. So what we demand by (9.18) or (9.23) is that the derived group does not coincide with the group; compare Section 6.3. And if we demand that we can repeat the procedure n times, we demand that the group is solvable (or integrable), that is, that the chain $G_r \supset G_{r'} \supset G_{r''} \supset \cdots \supset 1$, where $G_{r^{(i)}}$ is the derived group of $G_{r^{(i-1)}}$, ends with the identity (and the related chain of algebras with zero).

Collecting all the pieces, we can formulate the result as a theorem (due to Lie) as follows:

Lie's theorem. If an nth order differential equation admits a group G_n of n Lie point symmetries that is solvable and acts transitively in the space of first integrals, then the solution can be given in terms of n line integrals.

By the way, the groups were named solvable exactly because of this property!

If we have an nth order differential equation admitting a group G_r of $r \geqslant n$ symmetries, several cases may occur. We want to mention some of them.

If the group is not transitive, then a linear relation between $\mathbf{A}$ and the generators $\mathbf{X}_N$ exists, and the coefficients of the $\mathbf{X}_N$ in this relation may lead to first integrals; compare Section 8.2.

If the group G_r is solvable, then it contains (a) solvable subgroup(s) of order n. If one of these G_n acts transitively, we proceed as follows. We write down the commutators

$$[\mathbf{X}_N, \mathbf{X}_M] = C^P_{NM}\mathbf{X}_P, \qquad N, M, P = 1, \ldots, n, \tag{9.24}$$

and determine the set of $n - r'$ generators *not* spanned by the right side. If only one (represented by a linear combination of the given basic generators) exists, we take that as the new basic generator $\mathbf{X}_1$ and determine φ from (9.19). If there are several ($n - r' > 1$), then we introduce $n - r'$ new basic generators $\mathbf{X}_1, \ldots, \mathbf{X}_{n-r'}$ and apply (9.19) to each of them. We can now eliminate the $n - r'$ highest derivatives by means of the known $n - r'$ first integrals, choose an appropriate basis for the generators of the derived group $G_{r'}$, and start again by writing down the commutators for these generators.

If G_r is not solvable and does not contain a solvable subgroup G_n, then it may contain a subgroup G_n for which the process of taking the derived groups ends with a G_s, $s > 1$. We then can proceed as above to construct at least $n - s$ first integrals if G_n acts transitively.

9.4 Third integration strategy: differential invariants

If there is a solvable n-parameter group that acts transitively in the space of first integrals, then the second integration strategy provides the necessary means to solve the nth order differential equation. If there are less than n generators, or if the group is not solvable or not transitive, then we can use the first strategy and try to use the symmetries step by step. But here it is by no means sure that all symmetries can really be exploited since the reduced $(n-1)$th order differential equation need not admit the rest of the symmetries: equation (9.11) need not have $r-1$ solutions $\mathbf{X}_a$. Is there a more effective way to use the symmetries in these cases?

The answer to this question is provided by the following advice: find and use the differential invariants! As discussed in Section 6.4, differential invariants (of order k) of a group generated by r operators $\mathbf{X}_N$ are functions $\phi(x, y, y', \ldots, y^{(k)})$ that are invariant under the action of $\mathbf{X}_N$, that is, satisfy the r equations

$$\mathbf{X}_N \phi(x, y, y', \ldots, y^{(k)}) = 0. \tag{9.25}$$

If we compare this defining equation (9.25), with, for example, (5.18) and (9.13), then we see that we have already used some differential invariants in the other two integration strategies, although without mentioning this name. But if we are really to follow the advice given above, we have to know in some detail *how* to find and use differential invariants.

How can one find differential invariants? For an nth order differential equation we need all differential invariants of order $k \leqslant n$. As shown in Section 6.4, all differential invariants of arbitrary order can be obtained (by differentiation) from the two lowest order invariants φ and ψ. Thus finding the invariants essentially means finding φ and ψ. No algorithm is available to perform this task; we are in a similar situation to being told "transform a given generator to its normal form $\partial/\partial s$" and being promised it will work in many cases. Once we have found φ and ψ, we have to compute $\mathrm{d}\varphi/\mathrm{d}\psi$, $\mathrm{d}^2\varphi/\mathrm{d}\psi^2, \ldots$, until we have all invariants up to order n.

How can one use the differential invariants for the integration? Assume the maximum order of φ and ψ is i. We then take the list of differential invariants and the given differential equation $H(x, y, y', \ldots, y^{(n)}) = 0$ and express all derivatives $y^{(k)}$ with $k \geqslant i$ in terms of φ, ψ and the derivatives $\varphi^{(p)} = \mathrm{d}^p\varphi/\mathrm{d}\psi^p$, beginning with the highest order $y^{(n)}$. That will give an equation

$$\hat{H}(\psi, \varphi, \varphi', \ldots, \varphi^{(n-i)}) = 0. \tag{9.26}$$

The important point is that $\hat{H}$ will not depend explicitly on x, y, or derivatives of y essentially because it must be invariant under the group G_r and no invariants other than φ and ψ and the derivatives of φ with respect to ψ are available. We have thus reduced the given differential equation of order n to an equation of order $n-i$, which in general will not admit a Lie point symmetry (although in some cases the process of reduction may generate new symmetries; compare Section 12.1). Having found the general solution $\varphi = f(\psi)$ of (9.26), we write this solution as

$$\varphi(x, y, \ldots, y^{(i)}) = f[\psi(x, y, \ldots, y^{(c)})]. \tag{9.27}$$

This in fact is an ith order differential equation. It is clearly invariant under the r-parameter group G_r. The dimension p of the orbits of this group in $(x, y, y', \ldots, y^{(i)})$-space will be equal to i since it is an $(i+2)$-dimensional space and we know that exactly two invariants (φ and ψ) exist; compare Section 6.4 and equation (6.45). So (9.27) admits $r \geqslant i$ symmetries that can be used for its integration. In the general case the group will be neither solvable nor transitive in the space of first integrals so that it is not guaranteed that (9.27) can be solved by means of quadratures (line integrals), but quite often that will be possible.

The main advantage of using differential invariants is in the case when there are several, but not enough, symmetries: we then can split the task into the difficult part of integrating a differential equation (9.26) without symmetries and the easier part of integrating an equation (9.27) admitting comparatively many symmetries.

9.5 Examples

1. The third order differential equation

$$y''' = yy'' - y'^2 \tag{9.28}$$

admits the symmetries

$$\mathbf{X}_1 = \frac{\partial}{\partial x}, \qquad \mathbf{X}_2 = x\frac{\partial}{\partial x} - y\frac{\partial}{\partial y}, \qquad [\mathbf{X}_1, \mathbf{X}_2] = \mathbf{X}_1. \tag{9.29}$$

Since there are less than three symmetries, we try the first integration strategy discussed in Section 9.2. The commutator satisfies (9.12); that is, we can start (and must start) by transforming $\mathbf{X}_1$ – which in fact is unnecessary as it already has the desired form if we just exchange x and y. So with $s = x$, $t = y$, $s' = 1/y'$, and $s'' = -y''/y'^3$, we have

$$\begin{aligned}
\mathbf{A} &= \frac{\partial}{\partial t} + s'\frac{\partial}{\partial s} + s''\frac{\partial}{\partial s'} + \left(\frac{3s''^2}{s'} + s'^2 + ts's''\right)\frac{\partial}{\partial s''}, \\
\mathbf{X}_1 &= \frac{\partial}{\partial s} \\
\mathbf{X}_2 &= -t\frac{\partial}{\partial t} + s\frac{\partial}{\partial s} + 2s'\frac{\partial}{\partial s'} + 3s''\frac{\partial}{\partial s''},
\end{aligned} \tag{9.30}$$

that is, we have to deal with the second order (in s') differential equation

$$s''' = ts's'' + \frac{3s''^2}{s'} + s'^2, \qquad s' = \frac{ds}{dt}, \quad \text{etc.,} \tag{9.31}$$

and its symmetry

$$\mathbf{Y}_2 = -t\frac{\partial}{\partial t} + 2s'\frac{\partial}{\partial s''}. \tag{9.32}$$

We can now bring $\mathbf{Y}_2$ to its normal form by introducing coordinates u, v, which satisfy

$$\mathbf{Y}_2 u(t, s') = 1, \qquad \mathbf{Y}_2 v(t, s') = 0. \tag{9.33}$$

From (5.25)–(5.28) we learn that a possible solution is

$$v = s't^2, u = -\ln t \quad \Leftrightarrow \quad s' = ve^{2u}, t = e^{-u}. \tag{9.34}$$

In coordinates u and v, the differential equation (4.31) then reads

$$u'' = u'^3(v^2 - 6v) + u'^2(v - 7) - \frac{3u'}{v}, \qquad u' \equiv \frac{du}{dv}, \tag{9.35}$$

which reveals the rather nasty background of the innocent-looking differential equation (9.28): an Abelian differential equation for $u'(v)$. Having found $u'(v)$, we have to integrate it to get $u(v)$ and then use (9.34) to obtain $s'(t)$ and from it $s(t)$.

2. We will now try to treat the same differential equation (9.28) by using

the differential invariants of the generators:

$$\mathbf{X}_1 = \frac{\partial}{\partial x},$$

$$\mathbf{X}_2 = x\frac{\partial}{\partial x} - y\frac{\partial}{\partial y} - 2y'\frac{\partial}{\partial y'} - 3y''\frac{\partial}{\partial y''} - 4y'''\frac{\partial}{\partial y'''}. \tag{9.36}$$

Because of the structure of $\mathbf{X}_1$, differential invariants cannot depend on x. So the set of conditions (9.25) here reads

$$\left(y\frac{\partial}{\partial y} + 2y'\frac{\partial}{\partial y'} + 3y''\frac{\partial}{\partial y''} + 4y'''\frac{\partial}{\partial y'''}\right)\phi(y, y', y'', y''') = 0. \tag{9.37}$$

Invariants of order zero, $\phi = \phi(x, y)$, do not exist. A first order invariant is

$$\psi = y'/y^2 \tag{9.38}$$

or any function of it, and a second order invariant is provided by

$$\varphi = y''/y^3 \tag{9.39}$$

or any function of it (and of ψ). The third order invariant is then obtained from φ and ψ as

$$\varphi' = \frac{\mathrm{d}\varphi}{\mathrm{d}\psi} = \frac{y'''/y^3 - 3(y'y''/y^4)}{y''/y^2 - 2(y'^2/y^3)} = \frac{yy''' - 3y''y'}{y(yy'' - 2y'^2)}. \tag{9.40}$$

Eliminating from (9.28) all derivatives of y by means of (9.38)–(9.40), we obtain the first order (Abelian) differential equation for $\varphi(\psi)$:

$$(\varphi' - 1)(\varphi - 2\psi^2) + 3\varphi\psi - \psi^2 = 0. \tag{9.41}$$

From the solution $\varphi = f(\psi)$ of this equation, we get

$$y''/y^3 = f(y'/y^2), \tag{9.42}$$

a differential equation that can easily be solved using its two symmetries (9.36).

Comparing the two ways of treating the differential equation (9.28), one can say that the reduction to a first order equation is more straightforward when using differential invariants [note that the shape of this equation is different in the two approaches; compare (9.35) and (9.41)!]. On the other hand, when doing this reduction step by step as in the first approach, one already has to hand the details of the quadratures to be performed after solving that first order equation, which have to be added when really solving (9.42).

3. The differential equation

$$y''' = \frac{3}{2}\frac{y''^2}{y'} \tag{9.43}$$

admits the symmetries

$$\mathbf{X}_1 = \frac{\partial}{\partial y}, \qquad \mathbf{X}_2 = x\frac{\partial}{\partial x}, \qquad \mathbf{X}_3 = \frac{\partial}{\partial x}, \qquad \mathbf{X}_4 = y\frac{\partial}{\partial y}, \tag{9.44}$$

with commutators

$$\begin{aligned} &[\mathbf{X}_1, \mathbf{X}_2] = [\mathbf{X}_1, \mathbf{X}_3] = [\mathbf{X}_2, \mathbf{X}_4] = [\mathbf{X}_3, \mathbf{X}_4] = 0, \\ &[\mathbf{X}_1, \mathbf{X}_4] = \mathbf{X}_1, \qquad [\mathbf{X}_2, \mathbf{X}_3] = -\mathbf{X}_3. \end{aligned} \tag{9.45}$$

The first derived group has the generators $\mathbf{X}_1$ and $\mathbf{X}_3$, and the second derived group is the identity since $\mathbf{X}_1$ and $\mathbf{X}_3$ commute. So the above group G_4 is solvable. To apply the second integration strategy, we need a subgroup G_3. This group must contain the derived group (i.e., $\mathbf{X}_1$ and $\mathbf{X}_3$), and we choose $\mathbf{X}_2$ as the third generator. So we start from

$$\begin{aligned} \mathbf{X}_3 &= \frac{\partial}{\partial x}, \\ \mathbf{X}_1 &= \frac{\partial}{\partial y}, \\ \mathbf{X}_2 &= x\frac{\partial}{\partial x} - y'\frac{\partial}{\partial y'} - 2y''\frac{\partial}{\partial y''}, \\ \mathbf{A} &= \frac{\partial}{\partial x} + y'\frac{\partial}{\partial y} + y''\frac{\partial}{\partial y'} + \frac{3}{2}\frac{y''^2}{y'}\frac{\partial}{\partial y''}, \qquad \Delta = \frac{y''^2}{2}, \end{aligned} \tag{9.46}$$

where Δ is the determinant of this system. It does not vanish, so that the group G_3 acts transitively in the space of first integrals.

We now apply (9.19) with respect to generators $\mathbf{X}_1$ and $\mathbf{X}_2$ (which commute). We obtain

$$\varphi_1 = \int \frac{\begin{vmatrix} 1 & 0 & 0 & 0 \\ dx & dy & dy' & dy'' \\ x & 0 & -y' & -2y'' \\ 1 & y' & y'' & \frac{3}{2}(y''^2/y') \end{vmatrix}}{\Delta}$$

$$= \int \frac{\frac{1}{2}y''^2\,dy - 2y'y''\,dy' + y'^2\,dy''}{\frac{1}{2}y''^2} = y - \frac{2y'^2}{y''} \tag{9.47}$$

and

$$\varphi_2 = \ln y'^3 - \ln y''^2. \tag{9.48}$$

To substitute φ_1, φ_2 for y'', y', we write $\hat{\varphi}_2 = 4e^{\varphi_2}$ (any function of φ_2 is again a first integral) and get

$$y''^2 = 4y'^3/\hat{\varphi}_2, \qquad (y - \varphi_1)^2 = \hat{\varphi}_2 y'. \tag{9.49}$$

We need not go further within the general formalism, as it is clear that the last equation is the desired first order differential equation. It is easy to integrate and yields

$$y = \varphi_1 - \frac{\hat{\varphi}_2}{x - \varphi_3}. \tag{9.50}$$

[To allow for all possible limits $\varphi_i \to \infty$, $\varphi_i \to 0$, one could write this as $y = (ax + b)/(cx + d)$.]

9.6 Exercises

1 Prove that every nth order linear differential equation admits a solvable G_n.

2 Show that φ_3 in (9.39) satisfies $\mathbf{A}\varphi_3 = 0$.

3 Solve (9.43) by applying the first integration strategy.

4 Use the obvious symmetries of $4y^2y''' - 18yy'y'' + 15y'^3 = 0$ to solve it.

5 Prove that every G_2 is solvable.

6 Use the four symmetries $\mathbf{X}_1 = \partial/\partial x$, $\mathbf{X}_2 = \partial/\partial y$, $\mathbf{X}_3 = x\,\partial/\partial x + y\,\partial/\partial y$, $\mathbf{X}_4 = y\,\partial/\partial x - x\,\partial/\partial y$ to solve $y'''(1 + y'^2) = (3y' + a)y''^2$.

7 How is the differential invariant [solution to (9.37)] $\phi = y'''(y'')^{-4/3}$ related to the invariants (9.38)–(9.40)?

10

Systems of second order differential equations

10.1 The corresponding linear partial differential equation of first order and the symmetry conditions

Systems of second order differential equations most frequently occur in classical mechanics. We want to use the notation common in that field, so we take the time t as an independent variable and the generalized coordinates q^a as dependent variables and denote $\mathrm{d}q^a/\mathrm{d}t$ by $\dot{q}^a$. The system of second order differential equations ("equations of motion") we want to discuss then reads

$$\ddot{q}^a = \omega^a(q^i, \dot{q}^i, t), \qquad a, i = 1, \ldots, N. \tag{10.1}$$

The formal difference from the treatment of *one* differential equation can be diminished by again introducing *one* linear partial differential equation that is completely equivalent to the system (10.1). This partial differential equation is

$$\mathbf{A}f = \left(\frac{\partial}{\partial t} + \dot{q}^a\frac{\partial}{\partial q^a} + \omega^a(q^i, \dot{q}^i, t)\frac{\partial}{\partial \dot{q}^a}\right)f = 0 \tag{10.2}$$

(summation over the repeated index a). It admits $2N$ functionally independent solutions $\varphi^\alpha = \varphi^\alpha(q^a, \dot{q}^a, t)$ that are first integrals of the system (10.1).

To prove the equivalence of (10.1) and (10.2), we can follow exactly the reasoning presented in detail in Section 3.2 for *one* ordinary differential equation – only the number of variables and the detailed structure of $\mathbf{A}$ are different. So every solution of (10.1) can – locally – be written as $q^a =$

$q^a(\varphi^\alpha, t)$, where the φ^α are $2N$ constants of integration. From this we obtain $\varphi^\alpha = \varphi^\alpha(q^a, \dot{q}^a, t)$, and it follows from $\mathrm{d}\varphi^\alpha/\mathrm{d}t = 0$ and (10.1) that $\mathbf{A}\varphi^\alpha = 0$. Conversely, if we have $2N$ functionally independent solutions φ^α of $\mathbf{A}f = 0$ and compute $q^a = q^a(\varphi^\alpha, t)$ from $\varphi^\alpha(q^a, \dot{q}^a, t) = \text{const.}$, we have

$$\mathbf{A}\varphi^\alpha = \left(\frac{\partial}{\partial t} + \dot{q}^a \frac{\partial}{\partial q^a} + \omega^a \frac{\partial}{\partial \dot{q}^a}\right)\varphi^\alpha = 0,$$
$$\frac{\mathrm{d}\varphi^\alpha}{\mathrm{d}t} = \left(\frac{\partial}{\partial t} + \dot{q}^a \frac{\partial}{\partial q^a} + \ddot{q}^a \frac{\partial}{\partial \dot{q}^a}\right)\varphi^\alpha = 0. \tag{10.3}$$

These equations yield $(\omega^a - \ddot{q}^a)\,\partial\varphi^\alpha/\partial\dot{q}^a = 0$, and for functionally independent functions φ^α this is possible only if $\omega^a = \ddot{q}^a$.

By referring to the operator $\mathbf{A}$, most results valid for *one* (second order) differential equation can easily be generalized to *systems* of second order differential equations. A point transformation is now a (one-to-one) transformation between the variables q^a, t, and the generator of a group of point transformations is given by

$$\mathbf{X} = \xi(q^i, t)\frac{\partial}{\partial t} + \eta^a(q^i, t)\frac{\partial}{\partial q^a}. \tag{10.4}$$

The differential equations (10.1) admit the symmetry generated by the generator $\mathbf{X}$ and its extension

$$\dot{\mathbf{X}} = \xi\frac{\partial}{\partial t} + \eta^a\frac{\partial}{\partial q^a} + \dot{\eta}^a\frac{\partial}{\partial \dot{q}^a} \tag{10.5}$$

exactly if

$$[\dot{\mathbf{X}}, \mathbf{A}] = \lambda\mathbf{A} \tag{10.6}$$

holds (we shall drop the dot on $\dot{\mathbf{X}}$ if no confusion can arise from this). We need not determine the functions $\dot{\eta}^a$ from the extension of the finite group transformation but can consider the symmetry condition as defining them: the $\partial/\partial t$-component of (10.6) yields

$$-\mathbf{A}\xi = -\frac{\mathrm{d}\xi}{\mathrm{d}t} = \lambda, \tag{10.7}$$

and the $\partial/\partial q^a$-components then give

$$\dot{\eta}^a = \frac{\mathrm{d}\eta^a}{\mathrm{d}t} - \dot{q}^a\frac{\mathrm{d}\xi}{\mathrm{d}t}. \tag{10.8}$$

The essential part of the symmetry condition is contained in the $\partial/\partial\dot{q}^a$-components of (10.6). They read

$$\mathbf{X}\omega^a = \mathbf{A}\dot{\eta}^a - \omega^a\frac{\mathrm{d}\xi}{\mathrm{d}t}. \tag{10.9}$$

In full, this is (with $_{,t} = \partial/\partial t$, $_{,c} = \partial/\partial q^c$)

$$\begin{aligned}\xi\omega^a{}_{,t} + \eta^b\omega^a{}_{,b} + (\eta^b{}_{,t} + \dot{q}^c\eta^b{}_{,c} - \dot{q}^b\xi_{,t} - \dot{q}^b\dot{q}^c\xi_{,c})\frac{\partial\omega^a}{\partial\dot{q}^b}\\ + 2\omega^a(\xi_{,t} + \dot{q}^b\xi_{,b}) + \omega^b(\dot{q}^a\xi_{,b} - \eta^a{}_{,b}) + \dot{q}^a\dot{q}^b\dot{q}^c\xi_{,bc}\\ + 2\dot{q}^a\dot{q}^c\xi_{,tc} - \dot{q}^c\dot{q}^b\eta^a{}_{,bc} + \dot{q}^a\xi_{,tt} - 2\dot{q}^b\eta^a{}_{,tb} - \eta^a{}_{,tt} = 0.\end{aligned} \tag{10.10}$$

Equation (10.10) generalizes (4.11), which is contained as a special case. It serves to determine the components $\xi(q^i, t)$ and $\eta^a(q^i, t)$ of the symmetry generator $\mathbf{X}$ when ω^a (the set of differential equations) are given. The N equations ($a, b, c, \ldots = 1, \ldots, N$) are identities in t, q^i, and $\dot{q}^i$, and since ξ and η^a do *not* depend on $\dot{q}^i$, the N equations will split into many more equations and eventually make the integration possible.

If we have found one or several symmetries with generators $\mathbf{X}_N$, we can profit from the methods developed for one nth order differential equation since most of these do not rely on the explicit form of the operators $\mathbf{A}$ and $\mathbf{X}_M$ involved.

When there are less than $2N$ symmetries (less than the maximum number of first integrals), we can try to transform the symmetries step by step to a normal form in the space of variables (q^a, t) and thereby reduce the formal order $2N$ of the system; compare Section 9.2. For example if $\mathbf{X}_1 = \partial/\partial q_1$, so that the variable q_1 does not appear in any of the ω^a, the system is of first order with respect to $\dot{q}_1$. The actual transformation that gives, say, $\mathbf{X}_1$ its normal form may be more complicated to derive because of the larger number of variables involved; from

$$\mathbf{X}_1 = \xi_1(q^i, t)\frac{\partial}{\partial t} + \eta_1^a(q^i, t)\frac{\partial}{\partial q^a}, \tag{10.11}$$

we have to find functions s, t^k satisfying $\mathbf{X}_1 s = 1$, $\mathbf{X}_1 t^k = 0$. With some luck we might guess the correct solutions. Otherwise we should construct the trajectories (orbits) $q^i(t, q_0^i)$ of the group generated by $\mathbf{X}_1$ from

$$\frac{\mathrm{d}q^i}{\mathrm{d}\lambda} = \eta_1^i(q^a, t), \qquad \frac{\mathrm{d}t}{\mathrm{d}\lambda} = \xi_1(q^a, t), \tag{10.12}$$

that is, from

$$\frac{dq^1}{\eta_1^1}=\frac{dq^2}{\eta_1^2}=\cdots=\frac{dq^n}{\eta_1^n}=\frac{dt}{\xi_1}. \tag{10.13}$$

We can then take the (initial values) q_0^i as the t^i, use them as variables instead of, say, q^i, and then solve $\mathbf{X}_1 s = 1$ by a line integral; compare Section 5.2.

When the number of symmetries is at least $2N$, we can apply the results of Section 9.3; that is, for solvable groups we can use Lie's theorem and give all the necessary $2N$ first integrals in terms of line integrals.

We can also profit from using the differential invariants; compare Section 9.4. But although it remains true that if φ and ψ are differential invariants, so is $(d\varphi/dt)/(d\psi/dt)$, we shall not get all differential invariants by simply differentiating two of lowest order. So here (and at other places too) the procedure will be more involved than for just *one* second order equation.

In many physically important cases, the system $\ddot{q}^a = \omega^a$ can be derived from an action principle. In Section 10.3 we shall show how to use the Lagrangian to construct first integrals when the symmetries are known.

10.2 Example: the Kepler problem

In Cartesian coordinates q^a, the motion of a planet around the sun is governed by

$$\ddot{q}^a = \omega^a = -Mq^a/r^3, \qquad r^2 = q^i q_i, \qquad a, i = 1, 2, 3. \tag{10.14}$$

Since the functions ω^a do not depend on $\dot{q}^a$, it is rather easy to evaluate the symmetry conditions (10.10). The terms cubic and quadratic in $\dot{q}^a$ vanish only for

$$\xi = C_n(t)q^n + B(t), \qquad \eta^a = \tfrac{1}{2}(C_b\delta_c^a + C_c\delta_b^a)\dot{q}^b\dot{q}^c + D_b^a(t)q^b + E(t), \tag{10.15}$$

the linear terms vanish for $C_n = 0$, $\dot{D}_b^a = \frac{1}{2}\delta_b^a\ddot{B}$, and from the rest the functions ξ and η^a can be determined. We leave the details to the reader and state the final result, which is

$$\xi = b_1 t + b_2, \qquad \eta^a = \tfrac{2}{3}b_1 q^a + c_n\varepsilon^{na}{}_i q^i, \tag{10.16}$$

where b_1, b_2, c_1, c_2, c_3 are constants and ε^{nia} is completely antisymmetric with $\varepsilon^{123} = 1$.

The functions ξ and η^a contain five arbitrary parameters, so the equations of motion (10.14) admit a five-parameter group of Lie point symmetries with generators

$$\mathbf{X}_n = \varepsilon_n{}^k{}_a \left(q^a \frac{\partial}{\partial q^k} + \dot{q}^a \frac{\partial}{\partial \dot{q}^k} \right), \tag{10.17}$$

$$\mathbf{X}_4 = \frac{\partial}{\partial t}, \tag{10.18}$$

$$\mathbf{X}_5 = t\frac{\partial}{\partial t} + \tfrac{2}{3}q^a \frac{\partial}{\partial q^a} - \tfrac{1}{3}\dot{q}^a \frac{\partial}{\partial \dot{q}^a}. \tag{10.19}$$

The three generators $\mathbf{X}_n$ are those of the three-dimensional rotation group [compare (6.38) and Exercise 6.5]; they reflect the spherical symmetry of the gravitational field of the sun. The time translation $\mathbf{X}_4$ corresponds to the staticity of that field. But what is the physical meaning of the symmetry generated by $\mathbf{X}_5$? It is not even mentioned in most textbooks of classical mechanics! To study $\mathbf{X}_5$, one could, for example, construct a function ψ that satisfies $\mathbf{X}_5\psi = 0$ or ask for the finite transformation corresponding to $\mathbf{X}_5$. If one uses polar coordinates r, ϑ, φ instead of the Cartesian q^a, the generator $\mathbf{X}_5$ reads (without the extension to the derivatives)

$$\mathbf{X}_5 = t\frac{\partial}{\partial t} + \tfrac{2}{3}r\frac{\partial}{\partial r}, \tag{10.20}$$

and the corresponding finite transformations are

$$\tilde{t} = at, \qquad \tilde{r} = a^{2/3}r, \qquad a = \text{const.}, \tag{10.21}$$

from which

$$\tilde{t}^2/\tilde{r}^3 = t^2/r^3 \tag{10.22}$$

follows. If one scales a typical length of the planet's orbit and makes the appropriate change of its temporal behavior, then one again gets a possible orbit, and the two orbits are related by (10.22) – which is exactly Kepler's third law.

10.3 Systems possessing a Lagrangian: symmetries and conservation laws

From classical mechanics one knows that invariance under, for example, time translation (or rotation) implies conservation of energy (or angular momentum). This means that with each of the generators [e.g., $\mathbf{X}_4$

(or $\mathbf{X}_n$)] of the Kepler problem, one particular first integral can be associated. For an arbitrary system of N second order differential equations such a correspondence cannot be established unless the number of symmetries is $2N$. But it is possible to do this if the system is derivable from an action

$$W = \int_{t_1}^{t_2} L(q^k, \dot{q}^k, t)\,\mathrm{d}t, \tag{10.23}$$

a case that we study now.

We first define a Noether symmetry. A Noether symmetry is a Lie point transformation that leaves the action W invariant up to an additive constant $\hat{V}(\varepsilon)$, ε being the group parameter; this constant can be traced back to the time derivative of a function $\hat{V}(q^k, t, \varepsilon)$:

$$\tilde{W} = W + \hat{V}(\varepsilon) = W + \int_{t_1}^{t_2} \frac{\mathrm{d}\hat{V}(q^k, t, \varepsilon)}{\mathrm{d}t}\,\mathrm{d}t. \tag{10.24}$$

Since $\tilde{W}$ and W will lead to the same equations of motion (the same Euler–Lagrange equations), namely,

$$\frac{\mathrm{d}}{\mathrm{d}t}\frac{\partial L}{\partial \dot{q}^i} - \frac{\partial L}{\partial q^i} = 0, \tag{10.25}$$

Noether symmetries leave the differential equations invariant; they are a subset of the Lie point symmetries. If we denote the generator of a Noether symmetry by $\mathbf{X} = \xi\,\partial/\partial t + \eta^a\,\partial/\partial q^a + \dot{\eta}^a\,\partial/\partial \dot{q}_a$ and write down (10.24) in more detail, we obtain

$$\begin{aligned}\tilde{W} &= \int \tilde{L}\,\mathrm{d}\tilde{t} = \int L(\tilde{q}^k, \dot{\tilde{q}}^k, \tilde{t})\,\mathrm{d}\tilde{t}\\ &= \int [L(q^k, \dot{q}^k, t) + \varepsilon \mathbf{X}L + \cdots]\left(\mathrm{d}t + \varepsilon\frac{\mathrm{d}\xi}{\mathrm{d}t}\mathrm{d}t + \cdots\right)\\ &= \int L(q^k, \dot{q}^k, t)\,\mathrm{d}t + \varepsilon \int \frac{\mathrm{d}V}{\mathrm{d}t}\,\mathrm{d}t + \cdots.\end{aligned} \tag{10.26}$$

Comparing the terms of first order in ε and making use of the equations of motion by substituting $\mathrm{d}/\mathrm{d}t$ by the operator $\mathbf{A}$, we obtain from (10.26) the condition

$$\mathbf{X}L + (\mathbf{A}\xi)L = \mathbf{A}V, \qquad V = V(q^i, t), \tag{10.27}$$

for the existence of a Noether symmetry with generator **X**: if a function V satisfying (10.27) exists, then **X** is the generator of a Noether symmetry. Noether symmetries with $V=0$ are also called variational symmetries.

For the example of the Kepler problem in Section 10.2, the Lagrangian is

$$L=\frac{mM}{r}+\frac{m}{2}\dot{q}^a\dot{q}^a. \tag{10.28}$$

Of the five Lie point symmetries (10.17)–(10.18), the four $\mathbf{X}_n$ and $\mathbf{X}_4$ are Noether symmetries since

$$\mathbf{X}_n L=0, \qquad \mathbf{X}_4 L=\frac{\partial L}{\partial t}=0, \qquad \mathbf{A}\xi=0 \tag{10.29}$$

hold (V being zero for all of them), but $\mathbf{X}_5$ is not, as the right side of

$$\mathbf{X}_5 L+(\mathbf{A}t)L=\tfrac{1}{3}L \tag{10.30}$$

cannot be written as the time derivative of a function $V(q^k, t)$.

To establish the correspondence between Noether symmetries and first integrals, we shall prove that *if* $\mathbf{X}=\xi\partial/\partial t+\eta^a\partial/\partial q^a+\dot{\eta}^a\partial/\partial\dot{q}^a$ *is the generator of a Noether symmetry, then*

$$\varphi=\xi[\dot{q}^k L_{,\dot{q}^k}-L]-\eta^k L_{,\dot{q}^k}+V(q^i,t) \tag{10.31}$$

is a first integral that satisfies $\mathbf{X}\varphi=0$.

To show that φ is a first integral, we must prove that $\mathbf{A}\varphi=0$ holds. If we insert φ as given by (10.31) into $\mathbf{A}\varphi$ and make use of the fact that for solutions $\mathrm{d}/\mathrm{d}t$ can be replaced by **A**, so that

$$\frac{\mathrm{d}}{\mathrm{d}t}\frac{\partial L}{\partial\dot{q}_k}=\mathbf{A}L_{,\dot{q}^k}=L_{,q^k} \tag{10.32}$$

and

$$\dot{\eta}^a=\mathbf{A}\eta^a-\dot{q}^a\mathbf{A}\xi \tag{10.33}$$

hold, we obtain

$$\begin{aligned}\mathbf{A}\varphi=(\mathbf{A}\xi)[\dot{q}^k L_{,\dot{q}^k}-L]+\xi[\omega^a L_{,\dot{q}^a}+\dot{q}^k L_{,q^k}-L_{,t}-\dot{q}^a L_{,q^a}-\omega^a L_{,\dot{q}^a}]\\ -(\dot{\eta}^k+\dot{q}^k\mathbf{A}\xi)L_{,\dot{q}^k}-\eta^k L_{,q^k}+\mathbf{A}V,\end{aligned} \tag{10.34}$$

and if we replace $\mathbf{A}V$ by $\mathbf{X}L+L(\mathbf{A}\xi)$, we indeed have $\mathbf{A}\varphi=0$. The proof that $\mathbf{X}\varphi=0$ is also satisfied will be left to the reader.

Since in most cases it will be simple to determine V from (10.27) if the symmetry generator is known, equation (10.31) offers a straightforward way to get hold of a first integral. Of course, the condition (10.27) can also be used from the very beginning to determine the generators of any possible Noether symmetries, but one then may miss finding symmetries that are not Noether symmetries.

For the Kepler problem with Lagrangian (10.28) and the Noether symmetries (10.17) and (10.18), the first integrals computed by (10.31) are the angular momentum

$$\varphi_n = m\varepsilon_{nab}q^a\dot{q}^b, \tag{10.35}$$

and the energy

$$\varphi_4 = \frac{m}{2}\dot{q}^a\dot{q}^a - \frac{mM}{r}. \tag{10.36}$$

The fact that the first integral φ as given by equation (10.31) satisfies $\mathbf{X}\varphi = 0$ has an interesting consequence. Roughly speaking, the knowledge of a first integral amounts to the reduction of the order of the integration procedure by one (where for systems of N second order differential equations this order has to be counted as $2N$). If it is possible to replace one of the second order differential equations by the first order differential equation

$$\varphi(q^i, \dot{q}^i, t) = \varphi_0 = \text{const}, \tag{10.37}$$

then this differential equation $\varphi - \varphi_0 = 0$ admits the symmetry with generator $\mathbf{X}$ and can be integrated once by taking advantage of this symmetry. So this one Noether symmetry can lead to a reduction of order by two (see, e.g., the example given at the end of Section 5.2). It is due to this twofold reduction resulting from one symmetry that problems such as the Kepler problem (order 6) can be integrated despite an apparent shortage of symmetries.

10.4 Exercises

1 Show that the first integral φ defined by (10.31) satisfies $\mathbf{X}\varphi = 0$.

2 The Lagrangian for the harmonic oscillator is $L = \frac{1}{2}m\dot{q}^k\dot{q}^k - \frac{1}{2}Kq^kq^k$. Find all Lie point symmetries. Which of them are Noether symmetries? Construct the related first integrals.

3 Can one use the symmetry generator $\mathbf{X}_5$ of the Kepler problem to generate new first integrals from the known $\{\varphi_n, \varphi_4\}$ by applying it to them?

11

Symmetries more general than Lie point symmetries

11.1 Why generalize point transformations and symmetries?

Up to now we have considered only point transformations, that is, in the case of one dependent variable, transformations of the type

$$\tilde{y} = \tilde{y}(x, y; \varepsilon_N), \qquad \tilde{x} = \tilde{x}(x, y; \varepsilon_N), \tag{11.1}$$

and their extension

$$\tilde{y}' = \tilde{y}'(x, y, y'; \varepsilon_N), \qquad \tilde{y}'' = \tilde{y}''(x, y, y', y''; \varepsilon_N), \ldots, \tag{11.2}$$

to the higher derivatives, transformations that mix the dependent and the independent variable and may or may not depend on one or several parameters ε_N. If a given differential equation was invariant under (11.1) – now with at least one ε_N – or its infinitesimal generator

$$\mathbf{X}_N = \xi_N(x, y)\frac{\partial}{\partial x} + \eta_N(x, y)\frac{\partial}{\partial y} + \eta'_N(x, y, y')\frac{\partial}{\partial y'}$$

$$+ \eta''_N(x, y, y', y'')\frac{\partial}{\partial y''} + \cdots, \tag{11.3}$$

then we could determine this symmetry (if the differential equation was not of first order) and use it in the integration procedure.

Why should we generalize the allowed class of transformations and symmetries? Although we confined ourselves to Lie point *symmetries*, we in fact already used *transformations* more general than (11.1), and we met several results indicating that it may be worthwhile to enlarge the class of symmetries too. We suppressed these hints so far, but the reader may have noticed them. We want to name two of them now.

Suppose one has a second order differential equation that admits exactly one point symmetry with generator $\mathbf{X}_1$, as is the case for, for example,

$$y'' = \omega(y, y'), \qquad \mathbf{X}_1 = \frac{\partial}{\partial x}. \tag{11.4}$$

Then it is possible to reduce the differential equation to one of first order, which of course does not inherit any symmetry from the second order equation since none exists except that already used. But every first order equation admits a point symmetry (compare Section 4.2); why does this symmetry not show up in the original second order equation?

To answer this question, we write down the details of the procedure sketched above. The second order differential equation (11.4) becomes one of first order if we introduce $y' = v$ and $y = u$ as variables. We then have

$$\tilde{y} = v = y', \qquad \tilde{x} = u = y, \tag{11.5}$$

$$\tilde{y}' = \frac{d\tilde{y}}{d\tilde{x}} = \frac{dv}{du} = \frac{\omega(u, v)}{v}, \tag{11.6}$$

and the symmetry we know to exist is of the form

$$\mathbf{X}_2 = \xi(u, v)\frac{\partial}{\partial u} + \eta(u, v)\frac{\partial}{\partial v}. \tag{11.7}$$

In the original coordinates x, y this operator reads

$$\begin{aligned} \mathbf{X}_2 &= (\mathbf{X}_2 x)\frac{\partial}{\partial x} + (\mathbf{X}_2 y)\frac{\partial}{\partial y} + (\mathbf{X}_2 y')\frac{\partial}{\partial y'} \\ &= \xi(y, y')\frac{\partial}{\partial y} + \eta(y, y')\frac{\partial}{\partial y'}, \end{aligned} \tag{11.8}$$

so it is *not* a generator of a point transformation as the coefficient of $\partial/\partial y$ depends on y', and of course the transformation (11.5) and (11.6) is not a point transformation of type (11.1) and (11.2): $\mathbf{X}_2$ did not show up because it is not a *point* symmetry of $y'' = \omega(y, y'')$.

The second occasion when we met transformations that are not (extended) point transformations was when dealing with the so-called second integration strategy in Sections 7.3 and 9.3. Here the main idea was to use first integrals as new variables, and since first integrals of an nth order differential equation depend on x, y and all derivatives of y up to $y^{(n-1)}$, this amounts to making (nearly) arbitrary transformations in the space of variables $(x, y, y', \ldots, y^{(n-1)})$. Moreover, if, for example, we introduce for a second order differential equation two first integrals φ and ψ and a third function χ as variables instead of x, y, y', then because of

$$\mathbf{A}\varphi = \left(\frac{\partial}{\partial x} + y'\frac{\partial}{\partial y} + \omega(x, y, y')\frac{\partial}{\partial y'}\right)\varphi(x, y, y') = 0,$$

$$\mathbf{A}\psi(x, y, y') = 0, \qquad \mathbf{A}\chi(x, y, y') = 1, \tag{11.9}$$

in the new coordinates φ, ψ, χ, we have

$$\mathbf{A} = (\mathbf{A}\varphi)\frac{\partial}{\partial\varphi} + (\mathbf{A}\psi)\frac{\partial}{\partial\psi} + (\mathbf{A}\chi)\frac{\partial}{\partial\chi} = \frac{\partial}{\partial\chi}. \tag{11.10}$$

This operator $\mathbf{A}$ clearly admits the two "symmetries" $\mathbf{X}_1 = \partial/\partial\varphi$ and $\mathbf{X}_2 = \partial/\partial\psi$, in the sense of $[\mathbf{X}_1, \mathbf{A}] = 0 = [\mathbf{X}_2, \mathbf{A}]$, which of course are not generators of a point transformation group.

We hope that these few remarks will convince the reader that it makes sense to deal with transformations more general than point transformations and with the corresponding symmetries of differential equations. But one point must be stressed from the very beginning: if we enlarge the class of allowed transformations, more and more differential equations will possess symmetries, ending up with classes of transformations that contain invariance transformations for every differential equation. In the context of the integration procedure that means, unfortunately, that those symmetries will be less and less useful: we cannot, for example, expect to integrate every differential equation in terms of line integrals, and if every differential equation possesses a symmetry, we cannot expect that we will be able to find and use this symmetry.

11.2 How to generalize point transformations and symmetries

If one wants to generalize the transformations (11.1), then the examples given above indicate that one should mix and transform not only the independent and dependent variables x and y, but also the derivatives y', y'' up to a certain order p. A transformation that really puts variables and derivatives on an equal footing should read

$$\tilde{x} = \tilde{x}(x, y, y', y'', \ldots, y^{(p)}; \varepsilon_N),$$

$$\tilde{y} = \tilde{y}(x, y, y', y'', \ldots, y^{(p)}; \varepsilon_N), \tag{11.11}$$

and

$$\tilde{y}' = \tilde{y}'(x, y, y', y'', \ldots, y^{(p)}; \varepsilon_N),$$

$$\vdots$$

$$\tilde{y}^{(p)} = \tilde{y}^{(p)}(x, y, y', y'', \ldots, y^{(p)}; \varepsilon_N), \tag{11.12}$$

with arbitrary functions $\tilde{x}, \tilde{y}, \tilde{y}', \ldots, \tilde{y}^{(p)}$ (except, of course, that this transformation should be nonsingular).

If $y', y'', \ldots, y^{(p)}$ and their transforms were simply names for a certain set of variables, everything would be fine. But since they are derivatives, we have to insist on the conditions

$$\begin{aligned} \widetilde{y'} &= \tilde{y}' = \frac{\mathrm{d}\tilde{y}}{\mathrm{d}\tilde{x}}, \\ &\vdots \\ \widetilde{y^{(p)}} &= \widetilde{y^{(p-1)\prime}} = \frac{\mathrm{d}\tilde{y}^{(p-1)}}{\mathrm{d}x} \end{aligned} \tag{11.13}$$

(sometimes called the tangency conditions). These conditions imply that we are not free to choose the functions $\tilde{y}', \ldots, \tilde{y}^{(p)}$. If we have chosen $\tilde{x}$ and $\tilde{y}$, then everything is fixed; the equations (11.12) must be extensions of (11.11) to higher derivatives.

The difficulty we encounter is that the extensions of (11.1) do not have the form (11.12). That is, if $y^{(p)}$ is the highest derivative occurring in (11.11), then the extensions (11.12) will contain higher derivatives since every step forward in the extension will increase the order of the highest derivative appearing by one, and instead of (11.12) we will have

$$\begin{aligned} \tilde{y}' &= \tilde{y}'(x, y, y', \ldots, y^{(p+1)}; \varepsilon_N), \\ \tilde{y}^{(n)} &= \tilde{y}^{(n)}(x, y, y', \ldots, y^{(p+n)}; \varepsilon_N). \end{aligned} \tag{11.14}$$

There are two ways out of this apparent cul de sac. The first is to look for transformations that – although derivatives appear in (11.11) – by some miracle will not give rise to intolerable higher derivatives in the extensions. This idea, which leads to contact transformations, will be discussed in the following sections. The second possibility is to use the ordinary differential equation we want to study to eliminate all higher derivatives wherever they appear. This idea leads to the dynamical symmetries, to which Chapters 12 and 13 are devoted. In both cases we shall start from the infinitesimal generators and only occasionally refer to, and use, the finite transformations.

11.3 Contact transformations

The question we want to answer is the following: do generators

$$\mathbf{X} = \xi\frac{\partial}{\partial x} + \eta\frac{\partial}{\partial y} + \eta'\frac{\partial}{\partial y'} + \cdots + \eta^{(p)}\frac{\partial}{\partial y^{(p)}} \tag{11.15}$$

exist, where the coefficients $\xi, \eta, \eta', \ldots, \eta^{(p)}$ are functions of $x, y, y', \ldots, y^{(p)}$ but where the usual extension law (2.36), that is,

$$\eta^{(n)} = \frac{\mathrm{d}\eta^{(n-1)}}{\mathrm{d}x} - y^{(n)}\frac{\mathrm{d}\xi}{\mathrm{d}x}, \tag{11.16}$$

still holds? Transformations of this type are called contact transformations of order p (of the x–y plane).

The operator $\mathrm{d}/\mathrm{d}x$, if applied to functions depending on $y^{(p)}$, will generate terms with $y^{(p+1)}$. So the very essence of the question is whether these terms can cancel out. Since because of (11.16) we have

$$\eta^{(n)} = y^{(p+1)}\left[\frac{\partial\eta^{(n-1)}}{\partial y^{(p)}} - y^{(n)}\frac{\partial\xi}{\partial y^{(p)}}\right] + \text{terms without } y^{(p+1)}, \tag{11.17}$$

the terms in $y^{(p+1)}$ will cancel out only if

$$\frac{\partial\eta^{(n-1)}}{\partial y^{(p)}} = y^{(n)}\frac{\partial\xi}{\partial y^{(p)}}, \qquad 1 \leqslant n \leqslant p. \tag{11.18}$$

We must now check that this set of p equations does not contradict the extension law (11.16) with respect to the term *without* $y^{(p+1)}$. To do this, we write (11.16), for $n = p - 1$, as

$$\eta^{(p-1)} = \frac{\mathrm{d}}{\mathrm{d}x}(\eta^{(p-2)} - y^{(p-1)}\xi) + y^{(p)}\xi \tag{11.19}$$

and make use of the commutator relation (6.49) when inserting (11.19) into (11.18). We obtain

$$\begin{aligned}\frac{\partial\eta^{(p-1)}}{\partial y^{(p)}} &= \frac{\mathrm{d}}{\mathrm{d}x}\left(\frac{\partial\eta^{(p-2)}}{\partial y^{(p)}} - y^{(p-1)}\frac{\partial\xi}{\partial y^{(p)}}\right) + \frac{\partial\eta^{(p-2)}}{\partial y^{(p-1)}} \\ &\quad - y^{(p-1)}\frac{\partial\xi}{\partial y^{(p-1)}} + y^{(p)}\frac{\partial\xi}{\partial y^{(p)}} \\ &= y^{(p)}\frac{\partial\xi}{\partial y^{(p)}}\end{aligned} \tag{11.20}$$

and, making use of (11.18),

$$\frac{\partial \eta^{(p-2)}}{\partial y^{(p-1)}} = y^{(p-1)} \frac{\partial \xi}{\partial y^{(p-1)}}. \tag{11.21}$$

On the other hand, for $n = p - 1$, the equation

$$\frac{\partial \eta^{(p-2)}}{\partial y^{(p)}} = y^{(p-1)} \frac{\partial \xi}{\partial y^{(p)}} \tag{11.22}$$

follows from (11.18). The integrability condition $\partial^2 \eta^{(p-2)}/\partial y^{(p)} \partial y^{(p-1)} = \partial^2 \eta^{(p-2)}/\partial y^{(p-1)} \partial y^{(p)}$ of these two last equations yields $\partial \xi/\partial y^{(p)} = 0$, and if we insert this into (11.18), we get

$$\frac{\partial \xi}{\partial y^{(p)}} = 0, \qquad \frac{\partial \eta^{(n)}}{\partial y^{(p)}} = 0, \qquad 0 \leqslant n \leqslant p - 1. \tag{11.23}$$

This shows that if a contact transformation of order p exists at all, then all its coefficients ξ, $\eta^{(n)}$ are independent of the highest derivative $y^{(p)}$, with the possible exception of $\eta^{(p)}$. But that means that this contact transformation is in fact only of order $p - 1$ and has simply been extended to the pth derivative.

We now can repeat this reasoning and show that the maximum order is only $p - 2$, $p - 3$, and so on. This procedure breaks down when we arrive at $p = 1$, as then no coefficient $\eta^{(p-2)}$ as used in the proof exists, and we end up with the following result: if there are generators of contact transformations, then they are (extensions of)

$$\mathbf{X} = \xi(x, y, y')\frac{\partial}{\partial x} + \eta(x, y, y')\frac{\partial}{\partial y} + \eta'(x, y, y')\frac{\partial}{\partial y'}, \tag{11.24}$$

where because of (11.18)

$$\frac{\partial \eta}{\partial y'} = y' \frac{\partial \xi}{\partial y'} \tag{11.25}$$

holds.

The condition $\eta_{,y'} = y'\xi_{,y'}$ can easily be satisfied. If we introduce a function $\Omega(x, y, y')$ by $\xi = \Omega_{,y'}$, then the above condition reads $\eta_{,y'} = y'\Omega_{,y'y'}$, which is integrated by $\eta = y'\Omega_{,y'} - \Omega$. The question of whether contact transformations exist can now be answered in the positive:

Every infinitesimal contact transformation

$$\mathbf{X} = \xi(x, y, y')\frac{\partial}{\partial x} + \eta(x, y, y')\frac{\partial}{\partial y} + \eta'(x, y, y')\frac{\partial}{\partial y'} \tag{11.26a}$$

of the x–y plane can be given in terms of a generating function $\Omega(x, y, y')$ as

$$\xi = \frac{\partial \Omega}{\partial y'}, \qquad \eta = y'\frac{\partial \Omega}{\partial y'} - \Omega, \qquad \eta' = -\frac{\partial \Omega}{\partial x} - y'\frac{\partial \Omega}{\partial y}. \tag{11.26b}$$

If Ω is linear in y', then ξ and η do not depend on y': the transformation is a point transformation.

Why are these transformations called *contact* transformations. They are genuine transformations in the *three* variables x, y, y', so that in general the image of a curve $y = y(x)$ in (x, y)-space need not be a curve $\tilde{y} = \tilde{y}(\tilde{x})$ in (x, y)-space. But if two curves touch each other at a point, then this property is conserved in the sense that the images of the two curves also have a point $(\tilde{x}, \tilde{y})$ and a "direction" $\tilde{y}'$ in common.

We want to add a further result concerning contact transformations without going into the details: the *finite* contact transformation can also be given in terms of a generating function. If we have a function $V(x, y; \tilde{x}, \tilde{y})$ for which $V_{,x}(V_{,\tilde{x}}V_{,y\tilde{y}} - V_{,\tilde{y}}V_{,y\tilde{x}}) - V_{,y}(V_{,\tilde{x}}V_{,x\tilde{y}} - V_{\tilde{y}}V_{,x\tilde{x}})$ does not vanish, then one can resolve

$$V = 0, \qquad V_{,x} + y'V_{,y} = 0, \qquad V_{,\tilde{x}} + \tilde{y}'V_{,\tilde{y}} = 0, \tag{11.27}$$

with respect to either (x, y, y') or $(\tilde{x}, \tilde{y}, \tilde{y}')$ and obtain a contact transformation.

11.4 How to find and use contact symmetries of an ordinary differential equation

It is evident that we shall call a generator $\mathbf{X}$ of a contact transformation a generator of a contact *symmetry* if $\mathbf{X}$ leaves the ordinary differential equation $H(x, y, y', \ldots, y^{(n)}) = 0$ invariant, that is if

$$\mathbf{X}H = 0 \qquad (\mathrm{mod}\, H = 0) \tag{11.28}$$

or

$$[\mathbf{X}, \mathbf{A}] = \lambda \mathbf{A} \tag{11.29}$$

holds, where $\mathbf{A}$ is the linear operator equivalent to $H = 0$. The only difference from the point symmetries treated in the preceding chapters is that here the components ξ and η of the generator will depend on y'.

Is there an algorithmic way to determine any possible contact symmetries of a given ordinary differential equation? The answer to this question depends on the order of the differential equation.

For *first order* differential equations $y' = \omega(x, y)$, the symmetry condition (11.28) is essentially the same as for point symmetries since "mod $H = 0$" here means that y' should be replaced by $\omega(x, y)$ wherever it appears, that is, even in the generator, and we know from Section 4.2 that point symmetries always exist but they cannot practicably be found.

For *second order* differential equations $y'' = \omega(x, y, y')$, we have

$$\mathbf{A} = \frac{\partial}{\partial x} + y'\frac{\partial}{\partial y} + \omega\frac{\partial}{\partial y'}, \tag{11.30}$$

$$\mathbf{X} = \Omega_{,y'}\frac{\partial}{\partial x} + (y'\Omega_{,y'} - \Omega)\frac{\partial}{\partial y} - (\Omega_{,x} + y'\Omega_{,y})\frac{\partial}{\partial y'}, \tag{11.31}$$

and the symmetry condition $[\mathbf{X}, \mathbf{A}] = \lambda\mathbf{A}$ yields

$$\lambda = -\Omega_{,xy'} - y'\Omega_{,yy'} - \omega\Omega_{,y'y'} \tag{11.32}$$

and

$$\begin{aligned}\Omega_{,xx} &+ 2y'\Omega_{,xy} + y'^2\Omega_{,yy} + \omega(2\Omega_{,xy'} + \Omega_{,y} + 2y'\Omega_{,yy}) + \omega^2\Omega_{,y'y'} \\ &+ \omega_{,x}\Omega_{,y'} + \omega_{,y}(y'\Omega_{,y'} - \Omega) - \omega_{,y'}(\Omega_{,x} + y'\Omega_{,y}) = 0.\end{aligned} \tag{11.33}$$

This rather complicated equation (11.33) is the analogue of equation (4.11) valid for point symmetries. But in contrast to the case of point symmetries, it does *not* split into a set of differential equations since Ω depends on all three variables x, y, and y'. So in general it will be impossible to find a solution Ω to this equation although such solutions always exist. To prove this assertion, one could proceed as follows.

The general solution to $y'' = \omega(x, y, y')$ can always be written as $V_1(x, y; \varphi, \psi) = 0$, where φ and ψ are two (arbitrary) constants of integration. If we take φ and ψ as new variables $\tilde{x}$ and $\tilde{y}$, then V_1 can be chosen to be the generating function of a finite contact transformation [in the sense of (11.27)]. This particular contact transformation maps the solution curve with constants φ, ψ into the fixed points $\tilde{x} = \varphi$, $\tilde{y} = \psi$ (if one takes the initial values as coordinates, they remain constant along the solutions). The same procedure can be applied to the special second order differential equation $y'' = 0$, with $V_2(x, y; \varphi, \psi) = y - \varphi x - \psi$. Contact transformations form a group (which we did not show, but compare Exercise 12.2), and this means that if we first apply the contact transformation with generator V_1 and then the inverse transformation of that with generator V_2, we have mapped the (solutions of the) differential equation $y'' = \omega$ onto the (solutions of the) differential equation $y'' = 0$. The equation $y'' = 0$ admits contact symmetries, for example, the eight point symmetries (4.26) of the projective group, and so every second order differential equation admits contact symmetries.

So we can state that for second order differential equations the contact transformations play the same role as the point transformations do for first order differential equations: symmetries always exist but cannot be found explicitly, and all differential equations are equivalent with respect to the transformations (can be transformed into $y'' = 0$ or $y' = 0$, respectively).

With some luck, solutions to (11.33) can be found by some ad hoc *ansatz* for Ω, for example, by assuming that Ω has a special dependence (powers, polynomials, etc.) on one or some of its arguments: equation (11.33) will then split into several equations that may or may not have a nonzero solution. This, of course, is the counterpart of the attempts to find an integrating factor for a first order equation.

For *higher order* differential equations $y^{(n)} = \omega$, the symmetry condition (11.29),

$$\mathbf{X}\omega - \mathbf{A}\eta^{(n-1)} = -\omega(\mathbf{A}\Omega_{,y'}), \tag{11.34}$$

will be rather lengthy, but since Ω does not depend on $y'', y''', \ldots, y^{(n-1)}$, whereas ω and $\mathbf{A}$ do, this equation will split into several equations, and for arbitrary ω no solution Ω need exist; compare Exercise 11.4. Since convincing examples are not known (to us), we do not discuss more details but only mention Lie's result that for $n > 3$ all contact symmetries can be reduced to point symmetries by application of a suitable finite contact transformation to the differential equation and its symmetries.

Since we did not find contact transformations that are not point transformations and are not aware of applications, it does not make sense to enlarge upon how to use them. If one goes through the chapters dealing with applications of point symmetries, one will note that only a few technical things depend on the special form of the generators $\mathbf{X}$. So the main ideas remain undisturbed, in particular the procedure of trying to introduce normal forms of the generators in the space (now) of variables x, y, y' or in the space of first integrals.

11.5 Exercises

1 Usually contact transformations are defined as transformations $\tilde{x} = \tilde{x}(x, y, y')$, $\tilde{y} = \tilde{y}(x, y, y')$, $\tilde{y}' = \tilde{y}'(x, y, y')$ for which $dy - y'\,dx = \rho(\tilde{x}, \tilde{y}, \tilde{y}')\cdot(d\tilde{y} - \tilde{y}'\,dx)$ holds, with some nonzero function ρ. Show that this definition implies (11.25) for infinitesimal transformations.

2 Use the definition of Exercise 1 to formulate the condition(s) for contact transformations to form a group.

3 Give the explicit equations of the contact transformation generated by $V_2 = y - \tilde{x}x - \tilde{y}$.

4 Show that all contact symmetries of the third order differential equation $y''' = ax/y$, $a \neq 0$, are point symmetries, that is, Ω is necessarily linear in y'. [*Hint*: determine the coefficient of y'' in the symmetry condition (11.35).]

5 Find all contact symmetries of $y'' = 0$ with a generating function of the form $\Omega = x^n f(y, y')$.

12

Dynamical symmetries: the basic definitions and properties

12.1 What is a dynamical symmetry?

In nearly all preceding chapters we have made use of the fact that ordinary differential equations can be mapped to linear partial differential equations of first order. For an nth order differential equation $y^{(n)} = \omega(x, y, y', \ldots, y^{(n-1)})$ the corresponding linear operator $\mathbf{A}$ was

$$\mathbf{A} = \frac{\partial}{\partial x} + y'\frac{\partial}{\partial y} + \cdots + \omega\frac{\partial}{\partial y^{(n-1)}}, \tag{12.1}$$

for a system $\ddot{q}^a = \omega^a(q^i, \dot{q}^i, t)$ of second order equations we had

$$\mathbf{A} = \frac{\partial}{\partial t} + \dot{q}^a\frac{\partial}{\partial q^a} + \omega^a\frac{\partial}{\partial \dot{q}^a}, \tag{12.2}$$

and so on. To cover all cases by a uniform notation, we shall write the operator $\mathbf{A}$ as

$$\mathbf{A} = \frac{\partial}{\partial x} + a^k(x, z^i)\frac{\partial}{\partial z^k}, \qquad i, k = 1, \ldots, N, \tag{12.3}$$

where the variables z^i can be dependent variables or derivatives.

An operator

$$\mathbf{X} = \xi(x, z^i)\frac{\partial}{\partial x} + \eta^k(x, z^i)\frac{\partial}{\partial z^k} \tag{12.4}$$

is called a generator of a dynamical (or generalized) symmetry of the system represented by $\mathbf{A}$ as given by (12.3) exactly if

$$[\mathbf{X}, \mathbf{A}] = \lambda(x, z^i)\mathbf{A} \tag{12.5}$$

holds.

At first glance one may wonder why the components η^k of the generator (12.4) of a dynamical symmetry are not subject to some conditions that ensure that some of them are the extensions to higher derivatives of the basic components. But it is easy to see that all these conditions are contained in the set (12.5), which are in fact $N+1$ conditions that must be satisfied to ensure the existence of a symmetry; compare also Section 3.3.

In detail, the symmetry conditions read

$$-\mathbf{A}\xi = \lambda \tag{12.6}$$

for the $\partial/\partial x$-component, and if we insert this expression for λ back into (12.5), we obtain

$$\mathbf{X}a^k - \mathbf{A}\eta^k = -a^k\mathbf{A}\xi, \qquad k = 1, \ldots, N. \tag{12.7}$$

So if, for example, the operator $\mathbf{A}$ has the structure (12.1), then equations (12.7) read, for $k < n$,

$$\begin{aligned}\eta^{k+1} &= \mathbf{A}\eta^k - y^{(k)}\mathbf{A}\xi \\ &= \frac{\mathrm{d}\eta^k}{\mathrm{d}x} + [\omega - y^{(n)}]\frac{\partial \eta^k}{\partial y^{(n-1)}} - y^{(k)}\left\{\frac{\mathrm{d}\xi}{\mathrm{d}x} - [\omega - y^{(n)}]\frac{\partial \xi}{\partial y^{(n-1)}}\right\}.\end{aligned} \tag{12.8}$$

These equations coincide with the usual extension law

$$\eta^{k+1} = \frac{\mathrm{d}\eta^k}{\mathrm{d}x} - y^{(k)}\frac{\mathrm{d}\xi}{\mathrm{d}x} \tag{12.9}$$

if we just replace $y^{(n)}$ by ω wherever it appears. So the conditions for the existence of a dynamical symmetry contain the correct extension law and, moreover, give a precise meaning to the idea sketched in Section 12.3, namely, that it might be possible to admit derivatives in the components of a generator if we restrict the symmetry transformation to solutions of $y^{(n)} = \omega$.

The finite transformations associated with a given generator $\mathbf{X}$ of a dynamical symmetry can in principle be determined by integrating, for example, the set

$$\frac{\mathrm{d}\tilde{x}}{\mathrm{d}\varepsilon} = \xi(\tilde{x}, \tilde{y}, \tilde{y}', \tilde{y}'', \ldots, \tilde{y}^{(n-1)}),$$
$$\frac{\mathrm{d}\tilde{y}}{\mathrm{d}\varepsilon} = \eta(\cdots), \quad \frac{\mathrm{d}\tilde{y}'}{\mathrm{d}\varepsilon} = \eta'(\cdots), \ldots, \quad \frac{\mathrm{d}\tilde{y}^{(n-1)}}{\mathrm{d}\varepsilon} = \eta^{(n-1)}(\cdots) \tag{12.10}$$

of differential equations with initial values $\tilde{x} = x, \ldots, \tilde{y}^{(n-1)} = y^{(n-1)}$ for $\varepsilon = 0$. One will thus arrive at the transformations

$$\begin{aligned} \tilde{x} &= \tilde{x}(x, y, y', \ldots, y^{(n-1)}; \varepsilon), \\ \tilde{y} &= \tilde{y}(x, y, y', \ldots, y^{(n-1)}; \varepsilon), \\ \tilde{y}' &= \tilde{y}'(\cdots), \\ \tilde{y}^{(n-1)} &= \tilde{y}^{(n-1)}(\cdots), \end{aligned} \tag{12.11}$$

which could be called a "point" transformation in the $(x, y, y', \ldots, y^{(n-1)})$-space. Note, however, that $y^{(i+1)}$ will be the derivative $\mathrm{d}\tilde{y}^{(i)}/\mathrm{d}\tilde{x}$ only if the $\eta^{(k)}$ obey the correct extension law (12.9). Since in that extension law explicit use has been made of the differential equation $y^{(n)} = \omega(x, y, y', \ldots, y^{(n-1)})$, the finite transformation (12.11) makes sense and can be interpreted in terms of derivatives only for this special differential equation, or, stated otherwise, the equation $y^{(n)} = \omega$ emerges as an integrability condition of the system (12.11).

12.2 Examples of dynamical symmetries

Obviously, every point symmetry and every contact symmetry is also a dynamical symmetry [subject to the conditions that ξ and η are independent of the derivatives of y or that (11.26b) holds]. Interesting examples must be more general than these two classes.

For one first order differential equation,

$$\mathbf{A} = \frac{\partial}{\partial x} + \omega \frac{\partial}{\partial y}, \tag{12.12}$$

the possible generators $\mathbf{X}$ of a dynamical symmetry have the form

$$\mathbf{X} = \xi(x, y)\frac{\partial}{\partial x} + \eta(x, y)\frac{\partial}{\partial y}. \tag{12.13}$$

These are necessarily generators of a point symmetry. As we know from

Section 4.2, solutions to $[\mathbf{X}, \mathbf{A}] = \lambda \mathbf{A}$ always exist, but in most cases we shall be unable to find them.

For one second order differential equation,

$$\mathbf{A} = \frac{\partial}{\partial x} + y' \frac{\partial}{\partial y} + \omega(x, y, y') \frac{\partial}{\partial y'}, \tag{12.14}$$

the symmetry condition (12.5) for the generator,

$$\mathbf{X} = \xi(x, y, y') \frac{\partial}{\partial x} + \eta(x, y, y') \frac{\partial}{\partial y} + \eta'(x, y, y') \frac{\partial}{\partial y'}, \tag{12.15}$$

gives $\lambda = -\mathbf{A}\xi$,

$$\eta' = \mathbf{A}\eta - y'\mathbf{A}\xi \tag{12.16}$$

and

$$\xi\omega_{,x} + \eta\omega_{,y} + [(\mathbf{A}\eta) - y'(\mathbf{A}\xi)]\omega_{,y'} - \mathbf{A}(\mathbf{A}\eta) + y'\mathbf{A}(\mathbf{A}\xi) + 2\omega(\mathbf{A}\xi) = 0. \tag{12.17}$$

The function ω – and hence the operator $\mathbf{A}$ – being given, this is *one* linear partial differential equation for the *two* functions ξ and η. So solutions always exist, and one may even prescribe the function ξ, for example, by $\xi = 0$, but again no systematic way to find solutions is visible.

To give at least one simple example, we consider the differential equation

$$y'' = -y \tag{12.18}$$

of the harmonic oscillator. The symmetry condition (12.17) here reads

$$\eta + \mathbf{A}\mathbf{A}\eta - y'\mathbf{A}\mathbf{A}\xi + 2y\mathbf{A}\xi = 0, \qquad \mathbf{A} = \frac{\partial}{\partial x} + y' \frac{\partial}{\partial y} - y \frac{\partial}{\partial y'}. \tag{12.19}$$

Some simple solutions to these equations are

$$\xi = yy', \qquad \eta = -y^3, \tag{12.20}$$

$$\xi = y' \sin x, \qquad \eta = -y^2 \sin x, \tag{12.21}$$

$$\xi = y' \cos x, \qquad \eta = -y^2 \cos x. \tag{12.22}$$

Most applications and examples of dynamical symmetries are concerned with systems of second order differential equations derivable from a Lagrangian. We shall come back to these systems in Chapter 13.

12.3 The structure of the set of dynamical symmetries

As the examples given above indicate and as we shall show now in detail, there always exist infinitely many dynamical symmetries. Nevertheless, it is possible to give a clear picture of this manifold.

First, we observe that if **X** is a generator of a dynamical symmetry of a system characterized by the operator **A**, then

$$\hat{\mathbf{X}} = \mathbf{X} + \rho(x, z^k)\mathbf{A} \tag{12.23}$$

also represents a dynamical symmetry for arbitrary functions ρ. This is true since $[\hat{\mathbf{X}}, \mathbf{A}] = \lambda\mathbf{A} - (\mathbf{A}\rho)\mathbf{A} = \hat{\lambda}\mathbf{A}$ follows immediately from $[\mathbf{X}, \mathbf{A}] = \lambda\mathbf{A}$. To get rid of this rather uninteresting degree of freedom, we can introduce a certain gauge of the operators **X** by demanding that one particular component should be zero, for example, the coefficient ξ of $\partial/\partial x$. We then have

$$\mathbf{X} = \eta^k \frac{\partial}{\partial z^k}, \qquad \xi = 0, \tag{12.24}$$

and transformations (12.23) are no longer possible. Note that although the gauge (12.24) is always possible, it may not always be opportune to use it: a simple point transformation, with ξ and η only functions of, say, x and y,

$$\mathbf{X} = \xi(x, y)\frac{\partial}{\partial x} + \eta(x, y)\frac{\partial}{\partial y} + \cdots \tag{12.25}$$

would read

$$\mathbf{X} = (\eta - y'\xi)\frac{\partial}{\partial y} + \cdots \tag{12.26}$$

within this gauge, the coefficient of $\partial/\partial y$ now depending (linearly) on y'.

Second, we make use of the fact that an operator **A**, as given by (12.3), has exactly N functionally independent solutions (first integrals) φ^k satisfying

$$\mathbf{A}\varphi^k = 0, \qquad k = 1, \ldots, N. \tag{12.27}$$

If we use these φ^k as variables instead of z^k, then **A** and the possible symmetries **X** [in the gauge (12.24)] have the form

$$\mathbf{A} = \frac{\partial}{\partial x}, \tag{12.28}$$

$$\mathbf{X} = \phi^k(x, \varphi^i)\frac{\partial}{\partial \varphi^k}. \tag{12.29}$$

It is now easy to determine all dynamical symmetries: the symmetry condition $[\mathbf{X}, \mathbf{A}] = \lambda \mathbf{A}$ amounts to $\partial \phi^k/\partial x = 0$!

We can consider

$$\mathbf{X}_k = \frac{\partial}{\partial \varphi^k} \tag{12.30}$$

as constituting a basis for the (gauged) generators of dynamical symmetries, which of course depends on the choice of the set of first integrals φ^k. This basis has the property that all its elements commute with each other and with $\mathbf{A}$. The commutator of two arbitrary generators $\mathbf{X}_A$ and $\mathbf{X}_B$ (which are not necessarily elements of the basis) is again a dynamical symmetry, and thus a combination of the basis elements, but in general, with nonconstant coefficients,

$$[\mathbf{X}_A, \mathbf{X}_B] = C^k_{AB}(\varphi^i)\mathbf{X}_k, \tag{12.31}$$

the C^k_{AB} being functions of the first integrals.

The essence of these results remains true in arbitrary coordinates (x, z^i): there exist exactly N independent dynamical symmetries $\mathbf{X}_k$, and up to the addition of a nonconstant multiple of the operator $\mathbf{A}$, the general dynamical symmetry is a linear combination of these $\mathbf{X}_k$, the coefficients being arbitrary functions of the first integrals,

$$\mathbf{X} = \rho\mathbf{A} + \phi^k(\varphi^i)\mathbf{X}_k. \tag{12.32}$$

The commutator of two dynamical symmetries is a dynamical symmetry, but since the coefficients in (12.31) are not constant, the generators of the dynamical symmetry do *not* in general form a (finite-dimensional) Lie algebra.

The basis (12.30) of dynamical symmetries also suggests a simple interpretation of these symmetries: the general solution of an nth order differential equation can be written as

$$y = f(x; \varphi_1, \ldots, \varphi_n), \tag{12.33}$$

where the φ_i are constants of integration. Changing these constants will map solutions into solutions and thus be a symmetry operation, which of course looks complicated if the first integrals φ_i are written in terms of y and its derivatives.

12.4 Exercises

1 Show that the commutator of two generators of dynamical symmetries generates a dynamical symmetry.

2 Show that the dynamical symmetries (12.20)–(12.22) of the equation $y'' = -y$ can be obtained from the point symmetries $\mathbf{X}_1 = y\,\partial/\partial y$, $\mathbf{X}_2 = \sin x\,\partial/\partial y$, $\mathbf{X}_3 = \cos x\,\partial/\partial y$ by means of (12.32). Which functions ρ and which first integrals φ^i have been used?

3 Show that $\eta - y'\xi$ from (12.25) is invariant under transformations (12.23) of the generators.

4 Show that $\mathbf{X} = (x + y'/y)\,\partial/\partial x$ is a symmetry of $y''' = ayy''$, equivalent to a point symmetry.

5 Show that for $y'' = 0$ the general generator of a dynamical symmetry is (in the gauge $\xi = 0$) given by $\mathbf{X} = [g(y - xy', y') + xf(y - xy', y')]\,\partial/\partial y + \cdots$ with arbitrary f and g.

13

How to find and use dynamical symmetries for systems possessing a Lagrangian

13.1 Dynamical symmetries and conservation laws

Finding all dynamical symmetries of a given system of differential equations essentially means finding all of its first integrals and thus finding the general solution; this is one of the results of the analysis carried through in Section 12.3. So there is no hope of finding *all* dynamical symmetries in the general case, but with some luck we may find *some* of them if we make an appropriate *ansatz* for the components of the generator **X** (in fact, the situation is the complete analogue of the search for an integrating factor – or a point symmetry – for a first order differential equation). This has been done for several systems of second order differential equations that are derivable from a Lagrangian $L(q^a, \dot{q}^a, t)$ (compare also Section 10.3), and we shall concentrate on those systems from now on.

From Section 10.3 we know that if a Lagrangian exists, we could assign a first integral to every point symmetry that was also a Noether symmetry and that this first integral could be determined rather easily if the generator of the Noether symmetry was known. To make use of a dynamical symmetry, it would be very helpful if a similar procedure could be applied here.

Indeed, it is possible to define a subclass of the dynamical symmetries, the so-called Cartan symmetries, and to show that there is a one-to-one correspondence between gauged ($\xi = 0$) Cartan symmetries and first integrals φ and that one can construct the first integral once the Cartan symmetry is known. Whereas Noether symmetries are symmetries of the action principle (compare Section 10.3), which one can phrase as them being symmetries of the Noether form $L\,dt$, the Cartan symmetries are symmetries of the Cartan form $L\,dt + (dq^a - \dot{q}^a\,dt)\partial L/\partial \dot{q}^a = -H\,dt + p_a\,dq^a$ (H is the Hamiltonian and p_a the generalized momenta $\partial L/\partial \dot{q}_a$). We shall not go into the details of this interesting topic since for the purpose of finding the Cartan symmetries the results are rather meagre: they consist of advice on how to try to find them, and this advice can be understood without referring to the Cartan symmetries, although its background may remain unclear.

The only result of this theory of Cartan symmetries that we shall use is the following: if $L(q^a, \dot{q}^a, t)$ is the Lagrangian of the system characterized by

$$\mathbf{A} = \frac{\partial}{\partial t} + \dot{q}^a \frac{\partial}{\partial q^a} + \omega^a(q^k, \dot{q}^k, t)\frac{\partial}{\partial \dot{q}^a}, \qquad a, k = 1, \ldots, N, \tag{13.1}$$

and if

$$\mathbf{X} = \xi(q^k, \dot{q}^k, t)\frac{\partial}{\partial t} + \eta^a(q^k, \dot{q}^k, t)\frac{\partial}{\partial q^a} + \dot{\eta}^a(q^k, \dot{q}^k, t)\frac{\partial}{\partial \dot{q}^a} \tag{13.2}$$

is the generator of a Cartan symmetry (a special dynamical symmetry), then for $|\partial^2 L/\partial \dot{q}^a\,\partial \dot{q}^b| \neq 0$ there exists a first integral φ with $\mathbf{A}\varphi = 0 = \mathbf{X}\varphi$ such that

$$\frac{\partial^2 L}{\partial \dot{q}^a\,\partial \dot{q}^b}(\eta^a - \dot{q}^a \xi) = -\frac{\partial \varphi}{\partial \dot{q}^b} \tag{13.3}$$

holds [compare also (10.31)]. Conversely, if φ is a first integral, then $\eta^a - \dot{q}^a \xi$ as given by (13.3) determines a dynamical symmetry (up to a gauge transformation $\hat{\mathbf{X}} = \mathbf{X} + \rho\mathbf{A}$).

The reader need not bother about the difference between Cartan and dynamical symmetries since we shall use the equation (13.3) only in a rather pragmatic way that does not depend on this difference. Equation (13.3) obviously tells us that what counts most is the dependence of the Lagrangian L and the first integrals φ on the velocities $\dot{q}^a$. In most physical applications the Lagrangian will be quadratic (or a polynomial of order two) in the velocities $\dot{q}^a$, and it will make sense to look for first integrals that have a similar dependence on the $\dot{q}^a$.

So the result of all these lengthy considerations is the simple advice: if the Lagrangian is quadratic in the $\dot{q}^k$, look for first integrals φ (by solving $\mathbf{A}\varphi = 0$) that are polynomials in the $\dot{q}^k$. If first integrals φ have been found, one can determine the corresponding dynamical symmetry from (13.3) and thus obtain, perhaps, a better insight into the meaning and the properties of these first integrals.

13.2 Example: $L=(\dot{x}^2+\dot{y}^2)/2-a(2y^3+x^2y),\ a\neq 0$

To find first integrals, we start with the *ansatz*

$$\begin{aligned}\varphi &= \tfrac{1}{2}K_{ab}\dot{q}^a\dot{q}^b + K_a\dot{q}^a + K\\ &= \tfrac{1}{2}K_{11}\dot{x}^2 + K_{12}\dot{x}\dot{y} + \tfrac{1}{2}K_{22}\dot{y}^2 + K_1\dot{x} + K_2\dot{y} + K,\end{aligned} \tag{13.4}$$

where K_{ab}, K_a, and K are functions of x, y, and t to be determined from the condition that φ is indeed a first integral, that is, that $\mathrm{d}\varphi/\mathrm{d}t=0$ holds if the Euler–Lagrange equations

$$\ddot{x} = -2axy, \qquad \ddot{y} = -ax^2 - 6ay^2 \tag{13.5}$$

are satisfied.

In full the condition $\mathrm{d}\varphi/\mathrm{d}t=0$ reads

$$\begin{aligned}0 = {} & \tfrac{1}{2}K_{11,t}\dot{x}^2 + K_{12,t}\dot{x}\dot{y} + \tfrac{1}{2}K_{22,t}\dot{y}^2 + K_{1,t}\dot{x} + K_{2,t}\dot{y} + K_{,t}\\ & + \tfrac{1}{2}K_{11,x}\dot{x}^3 + K_{12,x}\dot{x}^2\dot{y} + \tfrac{1}{2}K_{22,x}\dot{x}\dot{y}^2 + K_{1,x}\dot{x}^2 + K_{2,x}\dot{x}\dot{y} + K_{,x}\dot{x}\\ & + \tfrac{1}{2}K_{11,y}\dot{x}^2\dot{y} + K_{12,y}\dot{x}\dot{y}^2 + \tfrac{1}{2}K_{22,y}\dot{y}^3 + K_{1,y}\dot{x}^2\dot{y} + K_{2,y}\dot{y}^2 + K_{,y}\dot{y}\\ & - 2axy(K_{11}\dot{x} + K_{12}\dot{y} + K_1) - (ax^2 + 6ay)(K_{12}\dot{x} + K_{22}\dot{y} + K_2).\end{aligned} \tag{13.6}$$

Since the K_{ab}, K_a, and K do *not* depend on $\dot{x}$ and $\dot{y}$, this condition splits into 10 equations. Equating to zero the coefficients of the terms of third order in the velocities $\dot{x}, \dot{y}$ yields

$$\begin{aligned}& K_{22,y} = 0 = K_{11,x}, \qquad \tfrac{1}{2}K_{22,x} + K_{12,y} = 0,\\ & \tfrac{1}{2}K_{11,y} + K_{12,x} = 0,\end{aligned} \tag{13.7}$$

which is integrated by

$$\begin{aligned}K_{11} &= \beta_1 + \beta_2 y + \beta_3 y^2,\\ K_{12} &= -\tfrac{1}{2}\beta_2 x - \beta_3 xy + \delta_1 y + \delta_2,\\ K_{22} &= \beta_3 x^2 - 2\delta_1 x + \alpha,\end{aligned} \tag{13.8}$$

where all coefficients may be functions of t. Inserting this back into (13.6), one will find that all the K_{ab}, K_a, and K are polynomials in x and y, and after some straightforward calculation one obtains finally

$$\begin{aligned}& K_{11} = \beta_1 + \beta_2 y, \qquad K_{12} = -\tfrac{1}{2}\beta_2 x,\\ & K_{22} = \beta_1, \qquad K_1 = 0, \qquad K_2 = 0, \qquad \beta_1, \beta_2 = \text{const.},\\ & K = a\beta_1(x^2y + 2y^3) - \tfrac{1}{8}a\beta_2(x^4 + 4x^2y^2).\end{aligned} \tag{13.9}$$

When going through the above routine, one will note that determining the coefficients K_{ab}, K_a, and K is very much the analogue of determining the point symmetries of a differential equation.

The appearance of two arbitrary constants β_1 and β_2 in (13.9) shows that we have in fact found two first integrals of the equations of motion (13.5), which are

$$\varphi_1 = \tfrac{1}{2}(\dot{x}^2 + \dot{y}^2) + a(x^2 y + 2y^3) \tag{13.10}$$

$$\varphi_2 = -\dot{x}^2 y + \dot{x}\dot{y}x + ax^2 y^2 + \tfrac{1}{4}ax^4. \tag{13.11}$$

We may now use equation (13.3) to determine the dynamical symmetries corresponding to these first integrals. It here reads, because $\partial^2 L/\partial \dot{q}^a \partial \dot{q}^b = \delta_{ab}$,

$$\eta^b - \dot{q}^b \xi = -\frac{\partial \varphi}{\partial \dot{q}^b}. \tag{13.12}$$

For φ_1, we obtain $\eta^b - \dot{q}^b \xi = -\dot{q}^b$, and we can choose $\xi = 1$, $\eta^b = 0$, that is, gauge the generator to be

$$\mathbf{X}_1 = \frac{\partial}{\partial t}. \tag{13.13}$$

This is of course a point symmetry, and since it expresses the invariance under time translation, the first integral φ_1 is to be called energy. For φ_2, we may choose $\xi = 0$ and then have

$$\mathbf{X}_2 = (2y\dot{x} - x\dot{y})\frac{\partial}{\partial x} - x\dot{x}\frac{\partial}{\partial y} + (\dot{x}\dot{y} + ax^3 + 2axy^2)\frac{\partial}{\partial \dot{x}} + (2ax^2 y - \dot{x}^2)\frac{\partial}{\partial \dot{y}}. \tag{13.14}$$

This is not a point symmetry since it is not possible to add a multiple of $\mathbf{A}$ and thereby to get rid of all derivatives in the coefficients of $\partial/\partial x$ and $\partial/\partial y$. No particular interpretation of this symmetry (and its first integral) is known.

The Lagrangian and the equations of motion admit a third symmetry,

$$\mathbf{X}_3 = t\frac{\partial}{\partial t} - 2x\frac{\partial}{\partial x} - 2y\frac{\partial}{\partial y} - 3\dot{x}\frac{\partial}{\partial \dot{x}} - 3\dot{y}\frac{\partial}{\partial \dot{y}}. \tag{13.15}$$

This again is a point symmetry (scaling symmetry). If we try to apply (13.13) to it, we see that $\partial\varphi/\partial\dot{q}^b$ should be linear in $\dot{q}^b$, so if there was a first integral φ corresponding to $\mathbf{X}_3$, it should be quadratic in the velocities and have shown up in the preceding *ansatz*; but $\mathbf{X}_3$ is neither a Noether nor a Cartan symmetry and thus is not related to a particular first integral.

The three symmetry generators have the commutators

$$[\mathbf{X}_1, \mathbf{X}_2] = 0, \qquad [\mathbf{X}_1, \mathbf{X}_3] = \mathbf{X}_1, \qquad [\mathbf{X}_2, \mathbf{X}_3] = 0, \tag{13.16}$$

that is, they happen to form a Lie algebra (which in general need not be the case).

13.3 Example: the Kepler problem

In Chapter 10, we already discussed the point symmetries of the Kepler problem with Lagrangian

$$L = \frac{m}{2}\dot{q}^a\dot{q}^a + \frac{mM}{r}, \qquad r^2 = q^a q^a, \qquad a = 1, \ldots, 3, \tag{13.17}$$

and found the first integrals corresponding to the four Noether symmetries among them. These first integrals were the three components of the angular momentum

$$\varphi^n = m\varepsilon^n{}_{ka} q^k \dot{q}^a, \qquad n, a, k = 1, \ldots, 3, \tag{13.18}$$

and the energy

$$E = \frac{m}{2}\dot{q}^a\dot{q}^a - \frac{mM}{r} \tag{13.19}$$

corresponding to the generators

$$\mathbf{X}^n = \varepsilon^{na}{}_k \left(q^k \frac{\partial}{\partial q^a} + \dot{q}^k \frac{\partial}{\partial \dot{q}^a} \right) \tag{13.20}$$

and

$$\mathbf{H} = -\dot{q}^a \frac{\partial}{\partial q^a} + \frac{M q^a}{r^3} \frac{\partial}{\partial \dot{q}^a}, \tag{13.21}$$

where we have added $-\mathbf{A}$ to $\mathbf{X}_4 = \partial/\partial t$ to obtain $\mathbf{H}$.

If we now ask for first integrals quadratic in the velocities $\dot{q}^k$,

$$\varphi = K_{ab}\dot{q}^a\dot{q}^b + K_a\dot{q}^a + K, \tag{13.22}$$

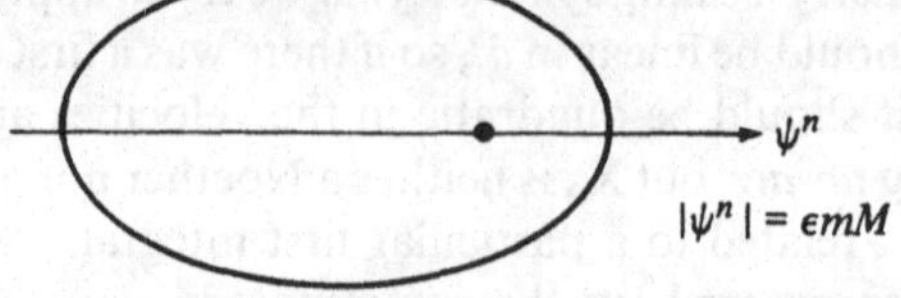

Figure 13.1. The Runge–Lenz-vector ψ^n for an elliptic orbit $r = k/(1 + \varepsilon \cos \varphi)$.

we shall of course regain the energy E but also the products $\varphi^n \varphi^k$ of components of the angular momentum. Are there still more? The answer is yes, and we shall simply state the result: there exists another vectorial first integral (conservation law), namely the Runge–Lenz vector (Figure 13.1)

$$\psi^n = mq^n(\dot{q}^a\dot{q}^a) - m\dot{q}^n(\dot{q}^a q^a) - m\frac{M}{r}q^n. \tag{13.23}$$

In terms of the planet's orbit, this conservation law says that the vector pointing from the sun ($q^n = 0$) to the nearest point of the orbit remains constant (there is no perihelion shift).

From (13.3) and in the gauge $\xi = 0$, we find the corresponding three dynamical symmetries to be

$$\mathbf{Y}^n = (\delta^{nk}\dot{q}^a q^a + \dot{q}^n q^k - 2q^n\dot{q}^k)\frac{\partial}{\partial q^k} + \left[\delta^{nk}\left(\dot{q}^a\dot{q}^a - \frac{M}{r}\right) - \dot{q}^n\dot{q}^k + \frac{M}{r^3}q^n q^k\right]\frac{\partial}{\partial \dot{q}^k}. \tag{13.24}$$

The commutators of the seven generators $\mathbf{H}$, $\mathbf{X}^n$, and $\mathbf{Y}^n$ turn out to be

$$\begin{aligned} &[\mathbf{X}^a, \mathbf{X}^b] = \varepsilon^{ab}{}_n\mathbf{X}^n, \quad [\mathbf{Y}^a, \mathbf{X}^b] = \varepsilon^{ab}{}_c\mathbf{Y}^c, \quad [\mathbf{H}, \mathbf{X}^a] = 0, \\ &[\mathbf{Y}^a, \mathbf{Y}^b] = -2E\varepsilon^{ab}{}_c\mathbf{X}^c + 2\varphi^c\varepsilon^{ab}{}_c\mathbf{H}, \quad [\mathbf{H}, \mathbf{Y}_a] = 0. \end{aligned} \tag{13.25}$$

The right sides are linear combinations of the generators, as was to be expected, but the coefficients are (nonconstant) functions of the first integrals (compare Section 12.3); the generators do not form a Lie algebra.

There is an interesting application of these generators in quantum mechanics, that is, for the hydrogen atom. If we consider bound states ($E < 0$), then we can introduce the generators

$$\mathbf{M}^a_\pm = \tfrac{1}{2}[\mathbf{X}^a \pm \mathbf{Y}^a/\sqrt{-2E}]. \tag{13.26}$$

Because of (13.25) and

$$\mathbf{X}E = \mathbf{Y}E = \mathbf{H}E = 0, \tag{13.27}$$

this set of six generators has the commutators

$$\begin{aligned} [\mathbf{M}^a_{(\pm)}, \mathbf{M}^b_{(\pm)}] &= \varepsilon^{ab}{}_c \mathbf{M}^c_{(\pm)} - \varepsilon^{ab}{}_c \frac{\varphi^c}{4E}\mathbf{H}, \\ [\mathbf{M}^a_+, \mathbf{M}^b_-] &= \varepsilon^{ab}{}_c \frac{\varphi^c}{4E}\mathbf{H}. \end{aligned} \tag{13.28}$$

If we now confine ourselves to properties for which we can set $\mathbf{H} \equiv 0$ (because of $\mathbf{H} \sim \partial/\partial t$ – properties that do not explicitly depend on time, such as the geometry of the orbits, or the quantum-mechanical eigenvalues), then we see that the system possesses two independent sets $\mathbf{M}^a_+$ and $\mathbf{M}^a_-$ of generators, both of the type of three-dimensional rotations, which now form a Lie algebra! In terms of this extra symmetry of the problem it can be understood that the energy eigenvalues depend only on the principal quantum number (only on the length of the major axis of the ellipse and not on the eccentricity). In Section 23.3 we shall come back to this example and show how these dynamical symmetries reenter as Lie–Bäcklund symmetries in the quantum-mechanical formulation of the Kepler problem.

13.4 Example: geodesics of a Riemannian space – Killing vectors and Killing tensors

In a Riemannian space with line element

$$\mathrm{d}s^2 = g_{ab}(x^n)\,\mathrm{d}x^a\,\mathrm{d}x^b, \qquad a, b, n = 1, \ldots, N, \tag{13.29}$$

the equation of the geodesics is derivable from the Lagrangian

$$L = \tfrac{1}{2} g_{ab} \frac{\mathrm{d}x^a}{\mathrm{d}s}\frac{\mathrm{d}x^b}{\mathrm{d}s} = \tfrac{1}{2} g_{ab}\dot{x}^a\dot{x}^b \tag{13.30}$$

(in this context one usually calls the coordinates x^a and not q^a). The corresponding Euler–Lagrange equations yield

$$\ddot{x}^a = -\Gamma^a_{bc}\dot{x}^b\dot{x}^c = -\tfrac{1}{2} g^{an}(g_{nb,c} + g_{nc,b} - g_{bc,n})\dot{x}^b\dot{x}^c, \tag{13.31}$$

where g^{an} is the inverse to g_{ab},

$$g^{an} g_{nb} = \delta^a_b \tag{13.32}$$

(we of course assume the symmetric matrix g_{ab} to be nonsingular), and Γ^a_{bc} are the Christoffel symbols, defined by (13.31). First integrals φ (which we hope to find) have, because of (13.31), to satisfy

$$\mathbf{A}\varphi = \left(\frac{\partial}{\partial s} + \dot{x}^a \frac{\partial}{\partial x^a} - \Gamma^n_{bc}\dot{x}^b\dot{x}^c \frac{\partial}{\partial \dot{x}^n}\right)\varphi = 0. \tag{13.33}$$

If we have found one, we can determine the associated dynamical symmetry from (13.3), which – after an inessential change of sign – here reads

$$g_{ab}(\eta^a - \xi\dot{x}^a) = \frac{\partial\varphi}{\partial\dot{x}^b}. \tag{13.34}$$

First integrals linear in $\dot{x}^a$ correspond to point symmetries in the gauge $\xi = 0$; the coefficients η^a do not depend on the $\dot{x}^a$! To find these first integrals, we start from

$$\varphi = K_a(x^i, s)\dot{x}^a + K(x^i, s) \tag{13.35}$$

and insert this into $\mathbf{A}\varphi = 0$. We thus obtain

$$\dot{x}^a \frac{\partial K_a}{\partial s} + K_{a,b}\dot{x}^a\dot{x}^b - \Gamma^c_{ab}\dot{x}^a\dot{x}^b K_c + \frac{\partial K}{\partial s} + K_{,a}\dot{x}^a = 0. \tag{13.36}$$

The coefficients of the different powers of x^a must vanish separately. If we use the covariant derivative defined by

$$K_{a;b} \equiv K_{a,b} - \Gamma^c_{ab}K_c, \tag{13.37}$$

this gives

$$K_{a;b} + K_{b;a} = 0 \tag{13.38}$$

and

$$K_{,a} + \frac{\partial K_a}{\partial s} = 0, \qquad \frac{\partial K}{\partial s} = 0. \tag{13.39}$$

Solutions K_a of (13.38) are called *Killing vectors* (usually one assumes that they do *not* depend on s). The associated generators

$$\mathbf{X} = g^{ab}K_b \frac{\partial}{\partial x^a} = K^a \frac{\partial}{\partial x^a} \tag{13.40}$$

correspond to the motions (translations and rotations) of a Euclidean space. Solutions K_a to (13.38) do not always exist, that is, not every Riemannian space has symmetries.

Equations (13.39) show that K_a is at most linear in s and that the coefficient of s must be a gradient:

$$K_a = \mathring{K}_a - sK_{,a}. \tag{13.41}$$

Because of (13.38), $\mathring{K}_a$ *and* $K_{,a}$ must be Killing vectors.

We now can summarize the result as follows. If a Riemannian space admits a Killing vector, that is, a solution $K_a(x^i)$ of (13.38), then

$$\varphi_1 = K_a\dot{x}^a \tag{13.42}$$

is a first integral. If this Killing vector is a gradient, $K_a = -K_{,a}$, then

$$\varphi_2 = K + sK_a\dot{x}^a \tag{13.43}$$

is also a first integral, and we have

$$K = \varphi_2 - s\varphi_1, \tag{13.44}$$

which is the analogue of, for example, $x = a + bt$ for a force-free motion in Euclidean space. This example shows that one point symmetry (Killing vector) can give two first integrals; compare Section 10.3.

First integrals quadratic in $\dot{x}^a$ have the form

$$\varphi = K_{ab}(x^i, s)\dot{x}^a\dot{x}^b + K_a(x^i, s)\dot{x}^a + K(x^i, s) \tag{13.45}$$

and must satisfy $\mathbf{A}\varphi = 0$, that is,

$$\begin{aligned}\dot{x}^a\dot{x}^b\frac{\partial K_{ab}}{\partial s} &+ \dot{x}^a\frac{\partial K_a}{\partial s} + \frac{\partial K}{\partial s} + K_{ab,c}\dot{x}^a\dot{x}^b\dot{x}^c\\ &+ K_{a,c}\dot{x}^a\dot{x}^c + K_{,c}\dot{x}^c - K_{ab}\dot{x}^b\Gamma^a_{nm}\dot{x}^n\dot{x}^m\\ &- K_{ab}\dot{x}^a\Gamma^b_{nm}\dot{x}^n\dot{x}^m - K_a\Gamma^a_{nm}\dot{x}^n\dot{x}^m = 0.\end{aligned} \tag{13.46}$$

Writing

$$K_{ab;c} \equiv K_{ab,c} - \Gamma^n_{bc}K_{na} - \Gamma^n_{ac}K_{nb}, \tag{13.47}$$

we obtain from the condition (13.46) the equations

$$K_{ab;c} + K_{bc;a} + K_{ac;b} = 0 \tag{13.48}$$

and

$$K_{a;b} + K_{b;a} + 2\frac{\partial K_{ab}}{\partial s} = 0, \quad K_{,a} + \frac{\partial K_a}{\partial s} = 0, \quad \frac{\partial K}{\partial s} = 0. \tag{13.49}$$

Symmetric tensors K_{ab} that are solutions to (13.48) are called *Killing tensors* of order two (usually only if independent of *s*). They correspond to true dynamical symmetries

$$\mathbf{X} = g^{an} K_{nb} \dot{x}^b \frac{\partial}{\partial x^a} + \cdots = K^a{}_b \dot{x}^b \frac{\partial}{\partial x^a} + \cdots. \tag{13.50}$$

From the system (13.49) we can infer the *s*-dependence of the various parts to be

$$\begin{aligned} K_{ab} &= \mathring{K}_{ab} - \tfrac{1}{2}s(\mathring{K}_{a;b} + \mathring{K}_{b;a}) + \tfrac{1}{2}s^2 K_{,a;b}, \\ K_a &= \mathring{K}_a - sK_{,a}. \end{aligned} \tag{13.51}$$

Those parts of K_a that do not contribute to K_{ab} are Killing vectors and lead to linear first integrals. We shall omit them here. Of course, each of the three parts of K_{ab} has to satisfy (13.48).

We can summarize the results as follows. If a Riemannian space admits a Killing tensor K_{ab}, then

$$\varphi_1 = K_{ab}\dot{x}^a\dot{x}^b \tag{13.52}$$

is a first integral. If this Killing tensor is derivable from a vector K_a according to $K_{ab} = -\tfrac{1}{2}(K_{a;b} + K_{b;a})$, then

$$\varphi_2 = -sK_{a;b}\dot{x}^a\dot{x}^b + K_a\dot{x}^a \tag{13.53}$$

is also a first integral, and we have

$$K_a\dot{x}^a = s\varphi_1 + \varphi_2. \tag{13.54}$$

If this vector K_a is a gradient, $K_{ab} = K_{,a;b}$, then there is a third first integral

$$\varphi_3 = \tfrac{1}{2}s^2 K_{,a;b}\dot{x}^a\dot{x}^b - sK_{,a}\dot{x}^a + K, \tag{13.55}$$

and we have

$$K = \tfrac{1}{2}s^2\varphi_1 + s\varphi_2 + \varphi_3. \tag{13.56}$$

One Killing tensor (one dynamical symmetry) may lead to three first integrals.

Killing vectors do not always exist. What about Killing tensors of order two? There always exists a trivial Killing tensor $K_{ab} = g_{ab}$, which leads to the first integral

$$\varphi = g_{ab}\dot{x}^a\dot{x}^b = 2L. \tag{13.57}$$

Even this trivial Killing tensor might possibly be expressible in terms of a vector:

$$g_{ab} = K_{a;b} + K_{b;a}. \tag{13.58}$$

If it satisfies this equation, K_a is called a *homothetic vector*. If there are Killing vectors $\underset{(\alpha)}{K_a}$, then

$$K_{ab} = A^{\alpha\beta}\left[\underset{(\alpha)}{K_a}\underset{(\beta)}{K_b} + \underset{(\beta)}{K_a}\underset{(\alpha)}{K_b}\right], \quad A^{\alpha\beta} = \text{const}, \tag{13.59}$$

is a Killing tensor, simply because any product of two linear first integrals yields a quadratic first integral. Like g_{ab}, this type of Killing tensor is called trivial. Examples of nontrivial Killing tensors are rare. There are none in Euclidean space. The most famous nontrivial example is that for the Kerr space-time (the gravitational field of a rotating black hole).

13.5 Exercises

1 Check that φ_2 as given by (13.11) satisfies $\mathbf{A}\varphi_2 = 0 = \mathbf{X}_2\varphi_2$ [with $\mathbf{X}_2$ from (13.14)].

2 Show that the seven first integrals φ^n, ψ^n, E of the Kepler problem cannot be independent, and find the connection between φ^n, E, and ψ^n. Can six of these first integrals be functionally independent? What is the consequence for the corresponding generators?

3 Show that (13.59) really defines a Killing tensor.

14

Systems of first order differential equations with a fundamental system of solutions

14.1 The problem

The subject we want to discuss in this chapter is not, strictly speaking, concerned with symmetries of a set of differential equations. But since the answer to the problem rests on the Lie algebra of certain generators and the problem itself is interesting enough to deserve being treated, we shall give in to this temptation.

One of the pleasant features of linear differential equations is that a superposition of solutions is possible. That is, it is always possible to obtain the general solution as a superposition of some special solutions multiplied by some arbitrary constants. For example, for the second order differential equation

$$y'' = f(x)y' + g(x)y + h(x) \tag{14.1}$$

one knows that the general solution can be expressed by a special solution y_{inh} of the inhomogeneous equation and two independent solutions y_1, y_2 of the homogeneous equation [(14.1) with $h = 0$] as

$$y = y_{\text{inh}} + \varphi^1 y_1 + \varphi^2 y_2 \tag{14.2}$$

φ^1 and φ^2 being arbitrary constants. If we write this as

$$y = \varphi^1(y_{\text{inh}} + y_1) + \varphi^2(y_{\text{inh}} + y_2) + (1 - \varphi^1 - \varphi^2)y_{\text{inh}}, \tag{14.3}$$

we see that we have given the general solution in terms of three special solutions y_{inh}, $y_{\mathrm{inh}} + y_1$, and $y_{\mathrm{inh}} + y_2$ and two arbitrary constants in the form of a relation,

$$y = F(y_A, \varphi^k), \qquad A = 1, 2, 3, \quad k = 1, 2. \tag{14.4}$$

Is it possible to generalize this property? Are there other classes of differential equations for which the general solution can be given in terms of some special solutions and a set of freely choosable constants, or – in other words – where a *nonlinear* superposition principle holds?

We shall give the answer to this question in the following section, only sketching the proof.

14.2 The answer

Consider the system

$$\frac{\mathrm{d}y^a}{\mathrm{d}x} = \omega^a(x, y^i), \qquad a, i = 1, \ldots, n, \tag{14.5}$$

of n ordinary differential equations of first order (which includes all nth order differential equations for one variable y if we set $y = y^1$, $y' = y^2$, $y'' = y^3, \ldots, y^{(n)} = \omega(x, y^1, \ldots, y^{(n-1)})$. Is it possible to write the general solution to this system in the form

$$y^a(x) = F^a[y^b_L(x), \varphi^k], \qquad a, b, c, k = 1, \ldots, n, \quad L = 1, \ldots, m, \tag{14.6}$$

where the $y^b_L(x)$ are special solutions of the system (14.5) and the φ^k are arbitrary constants? Note that x does not appear explicitly on the right side of (14.6)!

To derive a necessary condition, we use instead of the system (14.5) the corresponding partial differential equation

$$\mathbf{A}\varphi = \left(\frac{\partial}{\partial x} + \omega^a \frac{\partial}{\partial y^a}\right)\varphi = 0. \tag{14.7}$$

The n constants φ^k in (14.6) obviously correspond to the n independent solutions of (14.7); that is, if we solve the representation (14.6) of the solution for the φ^k,

$$\varphi^k = \varphi^k(y^a, y^b_L), \tag{14.8}$$

the functions φ^k have to satisfy $\mathbf{A}\varphi^k = \mathrm{d}\varphi^k/\mathrm{d}x = 0$ or

$$\mathbf{U}\varphi^k \equiv \left[\omega^a(x, y^b)\frac{\partial}{\partial y^a} + \omega^a_L(x, y^b_L)\frac{\partial}{\partial y^a_L} \right]\varphi^k = 0, \tag{14.9}$$

where the ω^a_L are the ω^a with y^b substituted by y^b_L. This is a partial differential equation in the $n(m+1)$ variables y^a and y^a_L.

As is obvious from (14.8), the φ^k do not explicitly depend on x, but in general the ω^a and ω^a_L do. So by assigning different fixed values x_σ to x, equation (14.9) is equivalent to many equations of the form

$$\mathbf{U}_\sigma\varphi^k = \left[\omega^a(x_\sigma, y^b)\frac{\partial}{\partial y^a} + \omega^a_L(x_\sigma, y^b_L)\frac{\partial}{\partial y^b_L} \right]\varphi^k = 0. \tag{14.10}$$

Of course, all commutators $[\mathbf{U}_\sigma, \mathbf{U}_\tau]$ and all linear combinations of the $\mathbf{U}_\sigma$ and their commutators will also annihilate the φ^k, and we end up with a system

$$\mathbf{V}_\sigma\varphi^k = 0, \qquad \sigma = 1, \ldots, r, \tag{14.11}$$

of partial differential equations the φ^k must satisfy. How many of these operators $\mathbf{V}_\sigma$ are independent of each other? In $n(m+1)$ variables, a system of r independent differential equations of that type will have $n(m+1) - r$ solutions, and since we assumed that n solutions φ^k exist, the number r of essentially different operators must be finite and limited by $r \leqslant n(m+1) - n$, that is, by $r \leqslant mn$. All $\mathbf{V}_\sigma$ arising from (14.10) must be linear combinations (not necessarily with constant coefficients) of a certain r-dimensional basic set. In particular, all commutators will be linear combinations of these basic elements,

$$[\mathbf{V}_\sigma, \mathbf{V}_\tau] = C^\lambda_{\sigma\tau}\mathbf{V}_\lambda. \tag{14.12}$$

Looking closer at the structure (14.10) of the $\mathbf{U}_\sigma$, which is also inherited by the $\mathbf{V}_\sigma$, we see that the coefficients of $\partial/\partial y^a$ do not depend on the y^a_L, and the coefficients of $\partial/\partial y^a_L$ (L fixed) do not depend on y^a and y^a_K ($L \neq K$). So the coefficients $C^\lambda_{\sigma\tau}$ cannot depend on either the y^a or the y^a_L; they must be constants, and the $\mathbf{V}_\sigma$ form a Lie algebra!

As already stated, we cannot obtain more than r operators $\mathbf{V}_\sigma$ by assigning special values x_σ to the x in equation (14.9). That means that the operator $\mathbf{U}$ defined by this equation, which in general *does* depend on x, must be a linear combination of the x-independent $\mathbf{V}_\sigma$, that is,

$$\mathbf{U} = \psi^\sigma(x)\mathbf{V}_\sigma = \psi^\sigma(x)\left[\eta^a_\sigma\frac{\partial}{\partial y^a} + \eta^a_{\sigma L}\frac{\partial}{\partial y^a_L} \right], \tag{14.13}$$

where the η^a_σ and $\eta^a_{\sigma L}$ are just names for the components of V_σ. Comparing (14.13) and (14.9), we see that $\omega^a = \psi^\sigma(x)\eta^a_\sigma$ must hold.

One can show that the condition just derived is also sufficient to guarantee the existence of a solution formula of type (14.6), and so we can summarize the results as follows.

A system of n ordinary differential equations

$$\frac{\mathrm{d}y^a}{\mathrm{d}x} = \omega^a(x, y^b), \qquad a, b = 1, \ldots, n, \tag{14.14}$$

of first order admits a fundamental system of m solutions $y^a_L(x)$ of (14.14), which allows the representation

$$y^a(x) = F^a[y^b_L(x), \varphi^k], \qquad a, b, k = 1, \ldots, n, \quad L = 1, \ldots, m, \tag{14.15}$$

of the general solution, exactly if there exist a set of r functions ψ^α and $n \cdot r$ functions $\eta^a_\alpha(y^b)$ such that

$$\omega^a = \psi^\alpha(x)\eta^a_\alpha(y^b), \qquad a, b = 1, \ldots, n, \quad \alpha = 1, \ldots, r, \tag{14.16}$$

holds and the r operators

$$\mathbf{X}_\alpha = \eta^a_\alpha \frac{\partial}{\partial y^a} \tag{14.17}$$

form a Lie algebra of dimension $r \leqslant m \cdot n$.

Unfortunately, no easy way to construct the functions F^a is known.

14.3 Examples

1. We know that a superposition principle holds for the general *linear* system

$$\frac{\mathrm{d}y^a}{\mathrm{d}x} = h^a_b(x)y^b + g^a(x) = \omega^a, \qquad a, b = 1, \ldots, n, \tag{14.18}$$

that is, a fundamental system of solutions exists. Of course, the functions $\psi^\alpha(x)$ are the functions $h^a_b(x)$ and $g^a(x)$, and the set of operators $\mathbf{X}_\alpha$ is given by

$$\mathbf{X}_\alpha = \left\{ y^b \frac{\partial}{\partial y^a}, \frac{\partial}{\partial y^c} \right\}, \qquad \alpha = 1, \ldots, r, \tag{14.19}$$

so that

$$\omega^a = \psi^\alpha \eta^a_\alpha. \tag{14.20}$$

It is easy to check that the $\mathbf{X}_\alpha$ indeed form a Lie algebra. In the general case, the number r of generators (14.19) is $n(n+1)$ because there are n^2 functions h^a_b and n functions g^a.

2. Suppose we have only one first order differential equation

$$\frac{\mathrm{d}y}{\mathrm{d}x} = \omega(x, y). \tag{14.21}$$

Which such first order equations have a fundamental system of solutions? Since only one dependent variable y occurs, the potential generators $\mathbf{X}_\alpha$ must be of the form

$$\mathbf{X}_\alpha = f_\alpha(y)\frac{\partial}{\partial y}, \tag{14.22}$$

and what we must find are all Lie algebras that possess a one-dimensional realization.

To do this, we start by transforming [by $y \to \phi(y)$], say, $\mathbf{X}_1$ to $\partial/\partial y$. We then have

$$\mathbf{X}_1 = \frac{\partial}{\partial y}, \qquad \mathbf{X}_\nu = f_\nu(y)\frac{\partial}{\partial y}, \qquad \nu = 2, \ldots, r. \tag{14.23}$$

Since all commutators $[\mathbf{X}_1, \mathbf{X}_\nu]$ belong to the Lie algebra, we know that all operators $\mathbf{X}^{(n)}_\nu = f^{(n)}_\nu \partial/\partial y$ ($f^{(n)} \equiv \mathrm{d}^n f/\mathrm{d}y^n$) are elements, and since there are at most r linearly independent elements, we conclude that each of the functions f_ν (ν fixed) admits a representation

$$f^{(r+1)}_\nu = c_\tau f^{(\tau)}_\nu, \qquad \tau = 1, \ldots, r, \tag{14.24}$$

that is, it is a solution of a linear differential equation, of order r at most, with constant coefficients c_τ. That means that all f_ν are analytic and so can be expressed as power series in y, and we have

$$\mathbf{X}_1 = \frac{\partial}{\partial y}, \qquad \mathbf{X}_\nu = y^{\lambda_\nu}(1 + a_\nu y + \cdots)\frac{\partial}{\partial y}. \tag{14.25}$$

The λ_ν are (positive) integers, and by a suitable choice of the basic generators, we can ensure that the integers λ_ν ($< r$) each appear only once, so that a "normal" basis is given by

$$\mathbf{X}_1 = \frac{\partial}{\partial y},$$

$$\mathbf{X}_2 = y(1 + a_2 y + \cdots)\frac{\partial}{\partial y},$$
$$\vdots \qquad (14.26)$$
$$\mathbf{X}_r = y^{r-1}(1 + a_r y + \cdots)\frac{\partial}{\partial y}.$$

The commutator of two arbitrary generators of this normal basis is again a generator. So for the commutator

$$[\mathbf{X}_{r-1}, \mathbf{X}_r] = y^{2r-4}(1 + \cdots)\frac{\partial}{\partial y} \qquad (14.27)$$

the exponent of y on the right side must be smaller than (or at most equal to) $r - 1$, that is, we have

$$r \leqslant 3. \qquad (14.28)$$

If a Lie algebra admits a one-dimensional realization, then its dimension r is at most three.

It is easy to show from the preceding that the most general Lie algebra with this property is given by

$$\mathbf{X}_1 = \frac{\partial}{\partial y}, \qquad \mathbf{X}_2 = y\frac{\partial}{\partial y}, \qquad \mathbf{X}_3 = y^2\frac{\partial}{\partial y}, \qquad (14.29)$$

and the two- or one-dimensional subalgebras contained here. By the way, the finite transformations corresponding to the algebra (14.29) are

$$\tilde{y} = \frac{a_1 + a_2 y}{a_3 + a_4 y}. \qquad (14.30)$$

We now come back to the question asked at the beginning. The answer obviously is: the most general first order differential equation that has a fundamental system of solutions is (in suitable coordinates y)

$$\frac{\mathrm{d}y}{\mathrm{d}x} = \psi^1(x) + \psi^2(x)y + \psi^3(x)y^2, \qquad (14.31)$$

that is, it is the Riccati differential equation (or one of its subcases). This particular property explains why the Riccati equation plays such an outstanding role among the first order differential equations.

The superposition law for the Riccati equation is well known. Because $r = 3$ and $n = 1$, we need at least $m = 3$ special solutions y_L, and three special solutions are also sufficient, as the formula

$$y = \frac{\varphi(y_1 - y_3)y_2 + (y_3 - y_2)y_1}{\varphi(y_1 - y_3) + (y_3 - y_2)} \tag{14.32}$$

shows.

14.4 Systems with a fundamental system of solutions and linear systems

The existence of a superposition law indicates that there is a close connection between linear systems and systems with a fundamental system of solutions. Suppose one has such a system,

$$\frac{dy^a}{dx} = \psi^\alpha(x)\eta^a_\alpha(y^i), \qquad i, a = 1, \ldots, n, \quad \alpha = 1, \ldots, r, \tag{14.33}$$

where

$$\mathbf{X}_\alpha = \eta^a_\alpha(y^i)\frac{\partial}{\partial y^a}, \qquad \alpha = 1, \ldots, r, \tag{14.34}$$

form a Lie algebra. Is it possible to transform this system into a linear one? The answer is yes, if "transformed" is correctly interpreted, and there are two possible approaches. Both rest on the fact that the Lie algebra of the $\mathbf{X}_\alpha$ is not changed when introducing new variables in the realization space or when this space is changed.

One knows from (14.18) and (14.19) that the generators $\mathbf{X}_\alpha$ of a linear system have η^a_α that are constant or linear in y^b. So the first approach is quite naturally to ask whether a transformation $\hat{y}^a = \hat{y}^a(y^b)$ exists that transforms the $\mathbf{X}_\alpha$ into this desired form. Or, stated differently, to ask whether a realization of the given Lie algebra in the given space exists that is linear. The answer can be yes or no, as two examples will illustrate.

The differential equation

$$\frac{dy}{dx} = \psi_1(x)y + \psi_2(x)y^2 \tag{14.35}$$

leads to the generators

$$\mathbf{X}_1 = y\frac{\partial}{\partial y}, \qquad \mathbf{X}_2 = y^2\frac{\partial}{\partial y}, \qquad [\mathbf{X}_1, \mathbf{X}_2] = \mathbf{X}_2. \tag{14.36}$$

The Lie algebra characterized by the commutator $[\mathbf{X}_1, \mathbf{X}_2] = \mathbf{X}_2$ admits a linear realization in a one-dimensional space: it is given by

$$\mathbf{X}_2 = \frac{\partial}{\partial \hat{y}}, \qquad \mathbf{X}_2 = -\hat{y}\frac{\partial}{\partial \hat{y}}, \tag{14.37}$$

and is obviously related to the original realization by $\hat{y} = -1/y$. In $\hat{y}$, the differential equation (14.35) reads

$$\frac{\mathrm{d}\hat{y}}{\mathrm{d}x} = -\psi_1(x)\hat{y} + \psi_2(x) \tag{14.38}$$

[it is not really necessary to transform (14.35); one simply has to replace the $\mathbf{X}_\alpha$ by their new realization (14.37)].

In contrast to this example, the Lie algebra

$$[\mathbf{X}_1, \mathbf{X}_2] = \mathbf{X}_1, \qquad [\mathbf{X}_1, \mathbf{X}_3] = 2\mathbf{X}_2, \qquad [\mathbf{X}_2, \mathbf{X}_3] = \mathbf{X}_3, \tag{14.39}$$

of the generators (14.29) appearing in the Riccati equation (14.31) does not admit a one-dimensional linear representation. So it is not possible to linearize the Riccati equation by a transformation $\hat{y} = \hat{y}(y)$.

The second approach is to ask for a linear representation of the given Lie algebra in a space of more than n dimensions. For the differential equations that means that we have to replace the original system in n (dependent) variables by a system in $p > n$ variables, which is linear. Those linear p-dimensional representations (with all the $\mathbf{X}$ nonzero) always exist.

For illustration, we take the Riccati equation (14.31) with the Lie algebra (14.38). The adjoint representation (6.44) gives a (intransitive) three-dimensional representation of this Lie algebra, but a two-dimensional representation also exists. It is given by

$$\mathbf{X}_1 = v\frac{\partial}{\partial u}, \quad \mathbf{X}_2 = \tfrac{1}{2}\left(u\frac{\partial}{\partial u} - v\frac{\partial}{\partial v}\right), \quad \mathbf{X}_3 = -u\frac{\partial}{\partial v}, \tag{14.40}$$

whereas the original nonlinear realization was

$$\mathbf{X}_1 = \frac{\partial}{\partial y}, \qquad \mathbf{X}_2 = y\frac{\partial}{\partial y}, \qquad \mathbf{X}_3 = y^2\frac{\partial}{\partial y}. \tag{14.41}$$

To find the connection between y and u, v, we apply (14.41) to the left side and (14.40) to the right side of $y = y(u, v)$. This yields

$$1 = v\frac{\partial y}{\partial u}, \qquad y = \frac{1}{2}\left(u\frac{\partial y}{\partial u} - v\frac{\partial y}{\partial v}\right), \qquad y^2 = -u\frac{\partial y}{\partial v}. \tag{14.42}$$

The unique solution to this system is easily found to be

$$y = u/v. \tag{14.43}$$

Before we start to write down the system of linear differential equations for the functions $u(x)$ and $v(x)$, we observe that in two variables there exist *four* operators $\mathbf{X}_\alpha$ with components linear in u and v. If the fourth operator $\mathbf{X}_4$ (which trivially forms a Lie algebra together with the other three – all of them are linear!) gives zero when projected into y-space, then it can be added in (u, v)-space without changing the Riccati equation in y-space. Indeed we can determine $\mathbf{X}_4$ from these conditions to be (up to a constant factor)

$$\mathbf{X}_4 = u\frac{\partial}{\partial u} + v\frac{\partial}{\partial v}, \qquad \mathbf{X}_4 y = 0, \qquad [\mathbf{X}_4, \mathbf{X}_\alpha] = 0. \tag{14.44}$$

Writing u for y^1, v for y^2, and λ for ψ^4 and applying equations (14.14), (14.16), and (14.17), we obtain

$$\begin{aligned} \frac{du}{dx} &= \psi^\alpha(x)\eta^1_\alpha(u, v) + \lambda(x)\eta^1_4(u, v), \\ \frac{dv}{dx} &= \psi^\alpha(x)\eta^2_\alpha(u, v) + \lambda(x)\eta^2_4(u, v), \qquad \alpha = 1, 2, 3. \end{aligned} \tag{14.45}$$

Inserting for η^a_α and η^a_4 their explicit expressions (14.40) and (14.44) yields

$$\begin{aligned} \frac{du}{dx} &= [\tfrac{1}{2}\psi^2(x) + \lambda(x)]u + \psi^1(x)v, \\ \frac{dv}{dx} &= -\psi^3(x)u \qquad + [\lambda(x) - \tfrac{1}{2}\psi^2(x)]v. \end{aligned} \tag{14.46}$$

This linear system (which is the most general homogeneous linear system in two variables) is completely equivalent to the Riccati equation

$$\frac{dy}{dx} = \psi^1(x) + \psi^2(x)y + \psi^3 y^2 \tag{14.47}$$

in the sense that every solution (u, v) of (14.46) gives a solution $y = u/v$ of (14.47), and if we split a solution y of (14.47) into u and v, then for every choice of u and v a function $\lambda(x)$ will exist such that (14.46) holds (the proof of the second part of this assertion is left to the reader).

14.5 Exercises

1 When are the $\mathbf{X}_\alpha$ of equation (14.17) symmetries of the system (14.14)? (Compare Exercise 4.7!)

2 Prove that for every splitting $y = u/v$ of a solution y of the Riccati equation (14.47) a function $\lambda(x)$ exists such that (14.46) holds.

II

Partial differential equations

15

Lie point transformations and symmetries

15.1 Introduction

Suppose we have a system

$$H_A(x^n, u^\alpha, u^\alpha_{,n}, u^\alpha_{,nm}, \ldots) = 0 \tag{15.1}$$

of partial differential equations in the N independent variables x^n, the M dependent variables $u^\alpha(x^n)$, and their derivatives, denoted by

$$u^\alpha_{,n} \equiv \frac{\partial u^\alpha}{\partial x^n}, \qquad u^\alpha_{,nm} \equiv \frac{\partial^2 u^\alpha}{\partial x^n \, \partial x^m}, \qquad \text{etc.} \tag{15.2}$$

What methods are available to solve those systems? In most cases the certainly frustrating answer will be that no practicable general recipes are known, at least if we mean by "solving" the construction of rather general solutions. In contrast to the case of ordinary differential equations, where the general solution will depend on arbitrary constants, we expect the general solution of a system of partial differential equations to depend on arbitrary functions of one or several variables, functions that reflect the freedom in the choice of initial or boundary values. But the solutions we are able to find will very often depend only on some parameters: they are rather special in character. Only in the case of linear differential equations can we use these special solutions to construct rather general classes of solutions by linear superposition.

Can symmetries help us to find at least such special solutions? The answer is an emphatic yes: most of the methods developed to find closed-form solutions of partial differential equations rest on symmetries, although this fact is not always stated explicitly when the method is presented.

To find these symmetries, we have to assume that they form a Lie group, so that we can use the infinitesimal generators instead of the finite group transformations in the search for them. As in the case of ordinary differential equations, we shall start with point transformations and point symmetries and only later deal with more general transformations and symmetries. Since much of what follows will be a repetition, only insignificantly enlarged, of what has been said in Chapter 2, we shall use here a rather concise form of presenting the fundamental definitions and properties of symmetries.

15.2 Point transformations and their generators

A point transformation is a (nonsingular) transformation between dependent and independent variables,

$$\begin{aligned} &\tilde{x}^n = \tilde{x}^n(x^i, u^\beta), && \tilde{u}^\alpha = \tilde{u}^\alpha(x^i, u^\beta), \\ &n, i = 1, \ldots, N, && \alpha, \beta = 1, \ldots, M. \end{aligned} \tag{15.3}$$

The introduction of new independent variables $\tilde{x}^n(x^i)$ or of new dependent variables $\tilde{u}^\alpha(u^\beta)$ is a quite common practice in dealing with partial differential equations. The complete mixing (15.3) of dependent and independent variables occurs less frequently.

As we shall see later, the symmetry transformations of partial differential equations can depend not only on one or several arbitrary parameters, but also on one or several arbitrary functions. In either case, a particular set of transformations

$$\tilde{x}^n = \tilde{x}^n(x^i, u^\beta; \varepsilon), \quad \tilde{u}^\alpha = \tilde{u}^\alpha(x^i, u^\beta; \varepsilon) \tag{15.4}$$

can be chosen that depend on only one parameter ε so that they form a group, with $\tilde{u}^\alpha = u^\alpha$ and $\tilde{x}^n = x^n$ for $\varepsilon = 0$. The corresponding infinitesimal transformation

$$\begin{aligned} &\tilde{x}^n = x^n + \varepsilon\xi^n(x^i, u^\beta) + \cdots, && \xi^n \equiv \left.\frac{\partial \tilde{x}^n}{\partial \varepsilon}\right|_{\varepsilon=0}, \\ &\tilde{u}^\alpha = u^\alpha + \varepsilon\eta^\alpha(x^i, u^\beta) + \cdots, && \eta^\alpha \equiv \left.\frac{\partial \tilde{u}^\alpha}{\partial \varepsilon}\right|_{\varepsilon=0}, \end{aligned} \tag{15.5}$$

is generated by

$$\mathbf{X} = \xi^n(x^i, u^\beta)\frac{\partial}{\partial x^n} + \eta^\alpha(x^i, u^\beta)\frac{\partial}{\partial u^\alpha}. \tag{15.6}$$

To apply the finite point transformation (15.4) and its generator (15.6) to a partial differential equation, we have to extend them to the derivatives. For the finite transformations, this is trivially done by defining

$$\tilde{u}^\alpha_{,n} = \frac{\partial \tilde{u}^\alpha}{\partial \tilde{x}^n}, \qquad \tilde{u}^\alpha_{,nm} = \frac{\partial^2 \tilde{u}^\alpha}{\partial \tilde{x}^n\, \partial \tilde{x}^m}, \quad \text{etc.} \tag{15.7}$$

To obtain the extension of the generator $\mathbf{X}$, we start from

$$\begin{aligned} \mathrm{d}\tilde{u}^\alpha &= \mathrm{d}u^\alpha + \varepsilon\, \mathrm{d}\eta^\alpha + \cdots \\ &= \left[\frac{\partial u^\alpha}{\partial x^i} + \varepsilon\left(\frac{\partial \eta^\alpha}{\partial x^i} + \frac{\partial \eta^\alpha}{\partial u^\beta}\frac{\partial u^\beta}{\partial x^i}\right)\right]\mathrm{d}x^i + \cdots, \end{aligned} \tag{15.8}$$

$$\begin{aligned} \mathrm{d}\tilde{x}^n &= \mathrm{d}x^n + \varepsilon\, \mathrm{d}\xi^n + \cdots \\ &= \left[\delta^n_m + \varepsilon\left(\frac{\partial \xi^n}{\partial x^m} + \frac{\partial \xi^n}{\partial u^\beta}\frac{\partial u^\beta}{\partial x^m}\right)\right]\mathrm{d}x^m + \cdots \end{aligned} \tag{15.9}$$

to calculate the functions η^α_n defined by

$$\tilde{u}^\alpha{}_{,n} = u^\alpha{}_{,n} + \varepsilon \eta^\alpha_n + \cdots, \qquad \eta^\alpha_n \equiv \left.\frac{\partial \tilde{u}^\alpha{}_{,n}}{\partial \varepsilon}\right|_{\varepsilon=0}. \tag{15.10}$$

In order to end up with a more transparent formula, we introduce the operator $\mathrm{D}/\mathrm{D}x^n$ by

$$\frac{\mathrm{D}}{\mathrm{D}x^n} = \frac{\partial}{\partial x^n} + u^\alpha{}_{,n}\frac{\partial}{\partial u^\alpha} + u^\alpha{}_{,nm}\frac{\partial}{\partial u^\alpha{}_{,m}} + \cdots. \tag{15.11}$$

Equations (15.8) and (15.9) can then be written as

$$\mathrm{d}\tilde{u}^\alpha = \left(\frac{\partial u^\alpha}{\partial x^i} + \varepsilon\frac{\mathrm{D}\eta^\alpha}{\mathrm{D}x^i}\right)\mathrm{d}x^i + \cdots, \tag{15.12}$$

$$\mathrm{d}\tilde{x}^n = \left(\delta^n_m + \varepsilon\frac{\mathrm{D}\xi^n}{\mathrm{D}x^m}\right)\mathrm{d}x^m + \cdots, \tag{15.13}$$

and we obtain from them

$$\frac{\partial \tilde{u}^\alpha}{\partial \tilde{x}^n} = \frac{u^\alpha_{,i} + \varepsilon \dfrac{D\eta^\alpha}{Dx^i} + \cdots}{\delta^n_m + \varepsilon \dfrac{D\xi^n}{Dx^m} + \cdots}\delta^i_m$$

$$= \left(u^\alpha_{,i} + \varepsilon \frac{D\eta^\alpha}{Dx^i} + \cdots\right)\left(\delta^i_n - \varepsilon \frac{D\xi^i}{Dx^n} + \cdots\right)$$

$$= u^\alpha_{,n} + \varepsilon \frac{D\eta^\alpha}{Dx^n} - \varepsilon u^\alpha_{,i}\frac{D\xi^i}{Dx^n} + \cdots. \tag{15.14}$$

Comparing this last equation with the definition (15.10) of the functions η^α_n, we read off the η^α_n as

$$\eta^\alpha_n = \frac{D\eta^\alpha}{Dx^n} - u^\alpha{}_{,i}\frac{D\xi^i}{Dx^n} = \frac{D}{Dx^n}(\eta^\alpha - u^\alpha{}_{,i}\xi^i) + \xi^i u^\alpha{}_{,in}. \tag{15.15}$$

So the generator (15.6), extended to the first derivatives, reads

$$\mathbf{X} = \xi^n(x^i, u^\beta)\frac{\partial}{\partial x^n} + \eta^\alpha(x^i, u^\beta)\frac{\partial}{\partial u^\alpha} + \eta^\alpha_n(x^i, u^\beta, u^\beta{}_{,i})\frac{\partial}{\partial u^\alpha{}_{,n}}, \tag{15.16}$$

with η^α_n given by (15.15).

In a similar way we can obtain the extensions to derivatives of arbitrary order, that is, the coefficients $\eta^\alpha_{nm}, \eta^\alpha_{nmp}, \ldots$ in

$$\mathbf{X} = \xi^n\frac{\partial}{\partial x^n} + \eta^\alpha\frac{\partial}{\partial u^\alpha} + \eta^\alpha_n\frac{\partial}{\partial u^\alpha{}_{,n}} + \eta^\alpha_{nm}\frac{\partial}{\partial u^\alpha{}_{,nm}} + \eta^\alpha_{nmp}\frac{\partial}{\partial u^\alpha{}_{,nmp}} + \cdots. \tag{15.17}$$

Starting, for example, from (15.13) and

$$d\tilde{u}^\alpha{}_{,n} = d(u^\alpha{}_{,n} + \varepsilon\eta^\alpha_n + \cdots) = \left(u^\alpha{}_{,nm} + \varepsilon\frac{D\eta^\alpha_n}{Dx^m} + \cdots\right)dx^m, \tag{15.18}$$

we get

$$\eta^\alpha_{nm} = \frac{D\eta^\alpha_n}{Dx^m} - u^\alpha_{,ni}\frac{D\xi^i}{Dx^m}, \tag{15.19}$$

which can also be written as

$$\eta^\alpha_{nm} = \frac{D}{Dx^m}(\eta^\alpha_n - u^\alpha{}_{,ni}\xi^i) + \xi^i u^\alpha{}_{,inm}$$

$$= \frac{D^2}{Dx^m Dx^n}(\eta^\alpha - u^\alpha{}_{,i}\xi^i) + \xi^i u^\alpha{}_{,imn}. \tag{15.20}$$

If we use the notation y^a to designate the variables x^i, u^α and all the partial derivatives (up to a fixed order, say),

$$y^a = \{x^i, u^\alpha, u^\alpha{}_{,i}, u^\alpha{}_{,ik}, \ldots\}, \tag{15.21}$$

then the generator $\mathbf{X}$ as given by (15.17) takes the form

$$\mathbf{X} = (\mathbf{X}y^a)\frac{\partial}{\partial y^a}. \tag{15.22}$$

This form is particularly useful when we perform a change

$$y^{a'} = y^{a'}(y^a) \tag{15.23}$$

of variables and want to express the generator $\mathbf{X}$ in terms of the new variables $y^{a'}$: we only have to calculate $\mathbf{X}y^{a'}(y^a)$ and write the resulting function in terms of the new variables to obtain the expressions sought as

$$\mathbf{X} = (\mathbf{X}y^{a'})\frac{\partial}{\partial y^{a'}}; \tag{15.24}$$

compare also Section 2.2.

As in the case of ordinary differential equations, transformations (15.23) that transform a given generator $\mathbf{X}$ to its normal form

$$\mathbf{X} = \frac{\partial}{\partial s} \tag{15.25}$$

always exist; that is, the system $\mathbf{X}y^{1'} = \mathbf{X}s = 1$, $\mathbf{X}y^{b'} = 0$ for $b' > 1$, always has a solution (which is not uniquely determined). For the generator of a point transformation it of course suffices to achieve this normal form (15.25) in the space of independent and dependent variables since if we have, for example, $\xi^1 = 1$, $\xi^n = 0 = \eta^\alpha$ ($n \neq 1$), then because of the extension formulae (15.15) and (15.19), $\eta^\alpha_n, \eta^\alpha_{nm}, \ldots$ vanish too.

15.3 The definition of a symmetry

A point symmetry of a set

$$H_A(y^a) = 0, \qquad y^a = \{x^i, u^\alpha, u^\alpha{}_{,i} u^\alpha{}_{,ik}, \ldots\} \tag{15.26}$$

of differential equations is a transformation

$$\tilde{y}^a = \tilde{y}^a(y^b; \varepsilon), \tag{15.27}$$

which is an extension to the derivatives of a point transformation (15.4) and which maps solutions into solutions. That is, if y^a is a solution of $H_A = 0$, then

$$H_A(\tilde{y}^a) = 0 \tag{15.28}$$

also holds for all values of ε.

By differentiating (15.28) with respect to ε, we obtain

$$\frac{\partial}{\partial \varepsilon} H_A(\tilde{y}^a)\bigg|_{\varepsilon=0} = \frac{\partial H_A}{\partial \tilde{y}^a}\frac{\partial \tilde{y}^a}{\partial \varepsilon}\bigg|_{\varepsilon=0} = \frac{\partial H_A}{\partial y^a}\frac{\partial \tilde{y}^a}{\partial \varepsilon}\bigg|_{\varepsilon=0} = 0, \tag{15.29}$$

and since the generator $\mathbf{X}$ of the transformation is defined by

$$\mathbf{X} = \frac{\partial \tilde{y}^a}{\partial \varepsilon}\bigg|_{\varepsilon=0}\frac{\partial}{\partial y^a}, \qquad \frac{\partial \tilde{y}^a}{\partial \varepsilon}\bigg|_{\varepsilon=0} = \{\xi^i, \eta^\alpha, \eta^\alpha_n, \eta^\alpha_{nm}, \ldots\}, \tag{15.30}$$

we can write (15.29) as

$$\mathbf{X}H_A = 0. \tag{15.31}$$

So if a point transformation (15.27) is a symmetry of the system $H_A = 0$, and if $H_A = 0$ holds, then $\mathbf{X}H_A = 0$ is valid.

To show that the converse is also true, one has to demand that the differential equations $H_A = 0$ have not been written in an inappropriate manner. If, for example, we were to write the two-dimensional potential equation as

$$H = (u_{,xx} + u_{,yy})^2 = 0, \tag{15.32}$$

then $\mathbf{X}H = 0$ would be valid for any linear operator $\mathbf{X}$ simply because *all* derivatives of H with respect to any of its arguments vanish on $H = 0$. To avoid such a singular behaviour of the equations $H_A = 0$, one has to impose the maximal rank condition

$$\operatorname{rank}\frac{\partial H_A}{\partial y^a} = Z \quad \text{on } H_A = 0, \tag{15.33}$$

where Z is the number of equations, $A = 1, \ldots, Z$. This condition ensures that not all derivatives of an H_A can vanish on $H_A = 0$ so that from $\mathbf{X}H_A = 0$

on $H_A = 0$ conclusions really can be drawn. In many applications the partial differential equations will be linear in the highest derivative, and the condition (15.33) will be satisfied trivially. We assume its validity from now on.

We leave the details of the proof (which runs in close analogy to that for ordinary differential equations) to the reader, and simply state the result: a Lie point transformation with generator **X** is a symmetry of a set $H_A = 0$ of partial differential equations exactly if

$$\mathbf{X}H_A \equiv 0 \qquad (\mathrm{mod}\, H_A = 0) \tag{15.34}$$

holds.

In (15.34), "mod $H_A = 0$" indicates that the differential equations $H_A = 0$ have to be used when evaluating $\mathbf{X}H_A = 0$, best done by eliminating as many of the highest derivatives as possible. If that has been done, then the symmetry condition (15.34) has to be satisfied *identically* in all remaining variables $(x^i, u^\alpha, u^\alpha_{,i}, \ldots)$ since it has to be true for every solution, which means that arbitrary values can be assigned to all these variables. In practice the symmetry condition (15.34) will often turn out to be the condition that $\mathbf{X}H_A$ is a linear combination of the H_A,

$$\mathbf{X}H_A = \lambda^B_A H_B \tag{15.35}$$

with some (nonconstant) coefficients λ^B_A.

It is important that the existence of a symmetry, that is, a solution **X** of the symmetry condition (15.34), does not depend on the choice of variables (x^i, u^α) in which the differential equations have been formulated. A change of variables may change the form of the differential equations (and of the components of the generator) considerably, but it will not touch the structure of the set of solutions to (15.34).

The question of finding (and then, of course, using) the symmetries will be treated in the following chapter(s).

15.4 Exercises

1 Show that the operators $\mathrm{D}/\mathrm{D}x^n$ and $\mathrm{D}/\mathrm{D}x^m$ [as defined by (15.11)] commute.

2 Show (by induction) that

$$\eta^\alpha_{n_1,\ldots,n_s} = \frac{\mathrm{D}^s}{\mathrm{D}x^{n_1}\mathrm{D}x^{n_2}\cdots\mathrm{D}x^{n_s}}(\eta^\alpha - u^\alpha{}_{,i}\xi^i) + \xi^i u^\alpha{}_{,in_1,\ldots,n_s}$$

holds.

3 Show that if (15.34) is valid, then the system $H_A = 0$ is invariant with respect to the finite transformations generated by **X**.

16

How to determine the point symmetries of partial differential equations

16.1 First order differential equations

Determining Lie point symmetries means finding functions $\xi^n(x^i, u^\beta)$ and $\eta^\alpha(x^i, u^\beta)$ such that the symmetry conditions $\mathbf{X}H_A = 0$ are satisfied. Compared with the case of ordinary differential equations, the number of functions will be larger and the explicit form of the symmetry conditions much more involved, but the general procedure remains the same: the symmetry conditions will lead to a system of *linear* partial differential equations for the ξ^n and η^α, which will split into many more equations since, for example, the ξ^n and η^α do not depend on the derivatives $u^\alpha{}_{,n}, u^\alpha{}_{,nm}, \ldots$, but the coefficients in the differential equations for the ξ^n and η^α do. In many cases, this system of equations can be solved straightforwardly.

The differential equations of a given set $H_A = 0$ need not be of the same order. However, we assume this in the examples and proceed from first to second order differential equations when illustrating the method in the following sections.

For first order differential equations, we need the generator $\mathbf{X}$ up to the first derivatives,

$$\mathbf{X} = \xi^n(x^i, u^\beta)\frac{\partial}{\partial x^n} + \eta^\alpha(x^i, u^\beta)\frac{\partial}{\partial u^\alpha} + \eta^\alpha_n\frac{\partial}{\partial u^\alpha{}_{,n}}, \tag{16.1}$$

where according to the extension law (15.15) the η^α_n are given by

$$\eta^\alpha_n = \frac{\mathrm{D}\eta^\alpha}{\mathrm{D}x^n} - u^\alpha{}_{,i}\frac{\mathrm{D}\xi^i}{\mathrm{D}x^n} = \eta^\alpha{}_{,n} + \eta^\alpha{}_{,\beta}u^\beta{}_{,n} - \xi^i{}_{,n}u^\alpha{}_{,i} - \xi^i{}_{,\beta}u^\beta{}_{,n}u^\alpha{}_{,i} \tag{16.2}$$

with

$$\eta^\alpha_{,n} \equiv \frac{\partial \eta^\alpha}{\partial x^n}, \qquad \eta^\alpha_{,\beta} \equiv \frac{\partial \eta^\alpha}{\partial u^\beta}, \quad \text{etc.} \tag{16.3}$$

We shall now study three relatively simple examples to see how the symmetry condition

$$\mathbf{X}H_A \equiv 0 \qquad (\operatorname{mod} H_A = 0) \tag{16.4}$$

can be solved in practice. When going through the calculations, the reader may perhaps not agree to call them simple – but solving nontrivial partial differential equations never is an easy task, and the same applies to finding their symmetries!

The *first example* is the general linear homogeneous differential equation

$$H = b^n(x^i)\varphi_{,n} = 0 \tag{16.5}$$

(where we have chosen to write φ instead of u^1). The partial differential equation

$$\mathbf{A}\varphi = \left(\frac{\partial}{\partial x} + y'\frac{\partial}{\partial y} + y''\frac{\partial}{\partial y'} + \cdots + \omega\frac{\partial}{\partial y^{(n)}}\right)\varphi = 0 \tag{16.6}$$

frequently used as the equivalent of an ordinary differential equation is contained here as a special case.

For equation (16.5), the condition $\mathbf{X}H \equiv 0$ amounts to

$$\xi^i b^n{}_{,i}\varphi_{,n} + (\eta_{,n} + \eta_{,\varphi}\varphi_{,n} - \xi^i{}_{,n}\varphi_{,i} - \xi^i{}_{,\varphi}\varphi_{,n}\varphi_{,i})b^n \equiv 0. \tag{16.7}$$

Now we must incorporate $H = b^n\varphi_{,n} = 0$ since $\mathbf{X}H \equiv 0$ is valid only $\operatorname{mod} H = 0$. This immediately yields

$$\xi^i b^n{}_{,i}\varphi_{,n} + \eta_{,n}b^n - b^n\xi^i{}_{,n}\varphi_{,i} \equiv 0. \tag{16.8}$$

The surviving terms with derivatives of φ must cancel out if $H = 0$ is taken into account since (16.8) has to be satisfied identically in x^i, φ, and $\varphi_{,n}$, and ξ^n and η do not depend on the derivatives $\varphi_{,n}$. This is possible only if

$$(\xi^i b^n{}_{,i} - b^i\xi^n{}_{,i})\varphi_{,n} = \lambda(x^m, \varphi)H = \lambda b^n\varphi_{,n} \tag{16.9}$$

holds with some function λ; compare (15.35). If we use the notation

$$\hat{\mathbf{X}} = \xi^i\frac{\partial}{\partial x^i}, \qquad \mathbf{B} = b^n\frac{\partial}{\partial x^n}, \tag{16.10}$$

then equation (16.9) reads

$$[\hat{\mathbf{X}}, \mathbf{B}] = \lambda \mathbf{B}, \tag{16.11}$$

which has exactly the form of the symmetry condition encountered in the context of ordinary differential equations; compare, for example, Sections 3.4, 13.1, and 13.2. The difference is that here – for the partial differential equation (16.6) – only those possible symmetries that have zero η are covered by $\hat{\mathbf{X}}$, and what is a point symmetry here corresponds to a dynamical symmetry of the ordinary differential equation.

We still have to satisfy the symmetry condition (16.8) for those terms that do not contain derivatives of φ. This leads of course to

$$b^n \eta_{,n} = 0, \tag{16.12}$$

which says that η is a solution of the differential equation $H = 0$. This symmetry corresponds to the possibility of adding an arbitrary solution to generate a new solution from a known one, which is a typical phenomenon occurring with linear differential equations.

To summarize, if we want to find all symmetries

$$\mathbf{X} = \xi^n(x^i, \varphi)\frac{\partial}{\partial x^n} + \eta(x^i, \varphi)\frac{\partial}{\partial \varphi} + \eta_n \frac{\partial}{\partial \varphi_{,n}} \tag{16.13}$$

of the homogeneous linear differential equation $H = b^n \varphi_{,n} = 0$, we have to determine ξ^n from (16.9) and η from (16.12). Solutions (i.e., symmetries) are known to exist, but in the general case we shall not be able to find them explicitly; compare also Chapter 12. This difficulty in really solving the symmetry condition is due to the linearity of the differential equation: it is not typical for first order differential equations.

The *second example* is the Hamilton–Jacobi equation

$$H = \tfrac{1}{2} g^{ab}(x^n) W_{,a} W_{,b} - E = 0 \tag{16.14}$$

in the form valid in curvilinear coordinates (and in general relativity), where E is a constant, the g^{ab} are the inverse of the matrix g_{ab} of the metric functions, and W is the function to be determined (the corresponding Lagrangian has been discussed in Section 13.4).

With

$$\mathbf{X} = \xi^i \frac{\partial}{\partial x^i} + \eta \frac{\partial}{\partial W} + (\eta_{,n} + \eta_{,W} W_{,n} - \xi^i{}_{,n} W_{,i} - \xi^i{}_{,W} W_{,n} W_{,i}) \frac{\partial}{\partial W_{,n}}, \tag{16.15}$$

the symmetry condition $\mathbf{X}H = 0$ reads

$$0 = \tfrac{1}{2}g^{ab}{}_{,n}\xi^n W_{,a} W_{,b} + (\eta_{,n} + \eta_{,W} W_{,n} - \xi^i{}_{,n} W_{,i} - \xi^i{}_{,W} W_{,n} W_{,i}) g^{an} W_{,a}. \qquad (16.16)$$

This condition has to be true for all solutions W of (16.14). If we use (16.14) and add a multiple of H, we obtain [with $\lambda = \lambda(x^i, W)$ still to be determined]

$$0 \equiv (\tfrac{1}{2}g^{ab}{}_{,n}\xi^n - \xi^b{}_{,n} g^{an}) W_{,a} W_{,b} + g^{an}\eta_{,n} W_{,a} + 2E\eta_{,W} - 2E\xi^i{}_{,W} W_{,i} + \lambda(g^{ab} W_{,a} W_{,b} - 2E) = 0. \qquad (16.17)$$

This must hold identically in all variables, and the coefficients of the different powers of $W_{,a}$ must vanish separately. So the symmetric part of the coefficient of $W_{,a} W_{,b}$ must be zero, which yields

$$\xi^n g_{ab,n} - \xi^b{}_{,n} g^{an} - \xi^a{}_{,n} g^{bn} = -2\lambda g^{ab}. \qquad (16.18)$$

We can write this equation in a more usual form if we introduce the quantities $\xi_n = g_{ni}\xi^i$, the Christoffel symbols Γ^a_{bc}, and the covariant derivative as in Section 13.4. It then reads

$$\xi_{a;b} + \xi_{b;a} = -2\lambda g_{ab}, \qquad \xi_n \equiv g_{ni}\xi^i. \qquad (16.19)$$

The rest of the symmetry condition (16.17) gives the two equations

$$g^{an}\eta_{,n} = 2E\xi^a{}_{,W}, \qquad (16.20)$$
$$(\eta_{,W} - \lambda)E = 0. \qquad (16.21)$$

Whether symmetries, that is, solutions $\xi^n(x^i, W)$, $\eta(x^i, W)$ to the system (16.19)–(16.21), exist depends entirely on the properties of the given functions $g^{ab}(x^i)$. If a solution $\xi_a(x^i)$ to (16.19) exists with a nonconstant $\lambda(x^i)$, then the space (or space-time) is said to admit a conformal motion or a conformal Killing vector, for constant λ a homothetic vector, and for $\lambda = 0$ a Killing vector.

To get a better understanding of the meaning of the three conditions (16.18), (16.20), and (16.21), we want to discuss the subcase $E \neq 0$, $\xi^a{}_{,W} = 0$ in more detail. In that case the three symmetry conditions imply

$$\eta = \eta(W), \qquad \xi^a = \xi^a(x^i), \qquad \eta_{,W} = \lambda(W), \qquad (16.22)$$
$$\xi^n g_{ab,n} - \xi^b{}_{,n} g^{an} - \xi^a{}_{,n} g^{bn} = -2\lambda g^{ab}. \qquad (16.23)$$

Since g_{ab} and ξ^a do not depend on W, the same is true for λ, so that, because of (16.22), λ must be constant and we have

$$\eta = \lambda W, \qquad \lambda = \text{const.} \tag{16.24}$$

(a possible additive constant η_0 has been neglected, as it corresponds to the trivial symmetry operation of adding a constant W_0 to a solution W of the Hamilton–Jacobi equation). Equation (16.23) – with $\lambda = \text{const.}$ – may or may not have a solution (the metric may or may not admit a homothetic vector). If a solution $\xi^n(x^i)$ exists, that is, if

$$\mathbf{X} = \xi^n(x^i)\frac{\partial}{\partial x^n} + \lambda W \frac{\partial}{\partial W} \tag{16.25}$$

is the generator of a symmetry, then it is always possible to perform a coordinate transformation (which also changes the functions g^{ab} to keep $H = 0$ invariant) such that

$$\xi^n = \delta^n_1 = (1, 0, 0, \ldots, 0) \tag{16.26}$$

holds. In these coordinates the symmetry condition (16.23) reads

$$\frac{\partial g^{ab}}{\partial x^1} = -2\lambda g^{ab}, \tag{16.27}$$

which is solved by

$$g^{ab} = \mathring{g}^{ab}(x^2, x^3, \ldots)e^{-2\lambda x_1}, \tag{16.28}$$

and the symmetry generator is

$$\mathbf{X} = \frac{\partial}{\partial x^1} + \lambda W \frac{\partial}{\partial W}. \tag{16.29}$$

So if a metric g^{ab} admits a homothetic motion (and the Hamilton–Jacobi equation the corresponding symmetry), the metric functions have a rather special dependence on the preferred coordinate x^1.

Really determining the symmetry amounts to solving the system (16.23) of partial differential equations for the functions ξ^n. Methods to do that have been developed, but we do not want to discuss them here.

The two examples just given show that for first order differential equations the task of finding the symmetries can be a rather complex one.

The *third example* is the system

$$H_1 = u_{,y} - v_{,x} = 0, \qquad H_2 = uu_{,x} + v_{,y} = 0 \tag{16.30}$$

of first order equations arising in gas dynamics.

To evaluate the symmetry conditions $\mathbf{X}H_A = 0$ with

$$\begin{aligned}\mathbf{X} = {} & \xi^1\frac{\partial}{\partial x} + \xi^2\frac{\partial}{\partial y} + \eta^1\frac{\partial}{\partial u} + \eta^2\frac{\partial}{\partial v} + \eta^1_x\frac{\partial}{\partial u_{,x}} \\ & + \eta^1_y\frac{\partial}{\partial u_{,y}} + \eta^2_x\frac{\partial}{\partial v_{,x}} + \eta^2_y\frac{\partial}{\partial v_{,y}},\end{aligned} \tag{16.31}$$

we first need the explicit expressions for $\eta^1_x, \eta^1_y, \ldots$. They are given by

$$\begin{aligned}\eta^1_x &= \eta^1{}_{,x} + u_{,x}(\eta^1{}_{,u} - \xi^1{}_{,x}) + u_{,y}(\eta^1{}_{,v} - \xi^2{}_{,x}) \\ &\quad - u^2_{,x}\xi^1{}_{,u} - u_{,x}u_{,y}(\xi^1{}_{,v} + \xi^2{}_{,u}) - u^2_{,y}\xi^2{}_{,v}, \\ \eta^1_y &= \eta^1{}_{,y} - u_{,x}(\xi^1{}_{,y} + u\eta^1{}_{,v}) + u_{,y}(\eta^1{}_{,u} - \xi^2{}_{,y}) \\ &\quad + u^2_{,x}u\xi^1{}_{,v} - u_{,x}u_{,y}(\xi^1{}_{,u} - u\xi^2{}_{,v}) - u^2_{,y}\xi^2{}_{,u}, \\ \eta^2_x &= \eta^2{}_{,x} + u_{,x}(\eta^2{}_{,u} + u\xi^2{}_{,x}) + u_{,y}(\eta^2{}_{,v} - \xi^1{}_{,x}) \\ &\quad + uu^2_{,x}\xi^2{}_{,u} - u_{,x}u_{,y}(\xi^1{}_{,u} - u\xi^2{}_{,v}) - u^2_{,y}\xi^1{}_{,v}, \\ \eta^2_y &= \eta^2{}_{,y} - uu_{,x}(\eta^2{}_{,v} - \xi^2{}_{,y}) + u_{,y}(\eta^2{}_{,u} - \xi^1{}_{,y}) \\ &\quad - u^2u^2_{,x}\xi^2{}_{,v} + uu_{,x}u_{,y}(\xi^1{}_{,v} + \xi^2{}_{,u}) - u^2_{,y}\xi^1{}_{,u},\end{aligned} \tag{16.32}$$

where all derivatives of v have been replaced by derivatives of u by means of the differential equations (16.30), so that in the symmetry conditions

$$\eta^1_y - \eta^2_x \equiv 0, \qquad \eta^1 u_{,x} + u\eta^1_x + \eta^2_y \equiv 0, \tag{16.33}$$

the equations $H_A = 0$ need not be taken into account (we also could have added multiples $\lambda_A H_A$ to the symmetry conditions and thus taken into account the "mod $H_A = 0$" in the original form of the symmetry conditions).

Inserting the explicit expressions for $\eta^1_x, \eta^1_y, \ldots$ into (16.33) and equating to zero the coefficients of the different powers of $u_{,x}$ and $u_{,y}$ [the equations (16.33) are also identities in these derivatives, and the functions ξ^1, ξ^2, η^1, and η^2 do not depend on them], we arrive at the system

$$\begin{aligned}&\eta^2{}_{,x} = \eta^1{}_{,y}, \qquad \eta^2{}_{,y} = -u\eta^1{}_{,x}, \\ &\eta^2{}_{,u} = -u\eta^1{}_{,v}, \qquad \eta^2{}_{,v} = \eta^1{}_{,u} + \eta^1/2u;\end{aligned} \tag{16.34}$$

$$\xi^1{}_{,x} = \xi^2{}_{,y} + \eta^1/2u, \qquad \xi^1{}_{,y} = -u\xi^2{}_{,x},$$
$$\xi^1{}_{,v} = \xi^2{}_{,u}, \qquad \xi^1{}_{,u} = -u\xi^2{}_{,v}. \tag{16.35}$$

It is easy to evaluate the integrability conditions (arising from, e.g., $\eta^2{}_{,xy} = \eta^2{}_{,yx}$) of these two sets of equations. They yield

$$2\xi^2{}_{,x} = -\eta^1{}_{,v}, \qquad u\eta^1{}_{,u} = \eta^1, \qquad \eta^1{}_{,x} = 0 = \eta^1{}_{,y}. \tag{16.36}$$

Starting from $\eta^1 = uf(v)$, it is a straightforward task to integrate the system (16.34) and (16.35). The result is that the general symmetry generator of the system $u_{,y} - v_{,x} = 0$, $uu_{,x} + v_{,y} = 0$ is given by

$$\begin{aligned} \mathbf{X} = {} & a_1\left[(u^2y - xv)\frac{\partial}{\partial x} - (ux + 2vy)\frac{\partial}{\partial y} + 2uv\frac{\partial}{\partial u} + (-\tfrac{2}{3}u^3 + \tfrac{3}{2}v^2)\frac{\partial}{\partial v}\right] \\ & + a_2\left[x\frac{\partial}{\partial x} + 2u\frac{\partial}{\partial u} + 3v\frac{\partial}{\partial v}\right] + a_3\frac{\partial}{\partial v} + a_4\left[x\frac{\partial}{\partial x} + y\frac{\partial}{\partial y}\right] \\ & + h(u,v)\frac{\partial}{\partial x} + g(u,v)\frac{\partial}{\partial y}, \end{aligned} \tag{16.37}$$

where $a_1, \ldots, a_4$ are arbitrary constants and the functions $h(u, v)$ and $g(u, v)$ are subject to the differential equations

$$h_{,u} = -ug_{,v}, \qquad h_{,v} = g_{,u}. \tag{16.38}$$

This result deserves two comments. The first is that we did not entirely succeed in determining the symmetries: we are left with the system (16.38) of partial differential equations which, although linear, cannot be solved in any generality. The second is that the symmetries divide into two sets: one depends on a finite number of arbitrary constants a_i and the other on an arbitrary function contained in the general solution of (16.38). This is a typical phenomenon often encountered with partial differential equations.

16.2 Second order differential equations

For second order differential equations, we need the generator **X** extended up to the second derivatives,

$$\begin{aligned} \mathbf{X} = {} & \xi^n(x^i, u^\beta)\frac{\partial}{\partial x^n} + \eta^\alpha(x^i, u^\beta)\frac{\partial}{\partial u^\alpha} \\ & + \eta^\alpha_n\frac{\partial}{\partial u^\alpha{}_{,n}} + \eta^\alpha_{nm}\frac{\partial}{\partial u^\alpha{}_{,nm}}, \end{aligned} \tag{16.39}$$

with

$$\eta^\alpha_n = \eta^\alpha{}_{,n} + \eta^\alpha{}_{,\beta} u^\beta{}_{,n} - \xi^i{}_{,n} u^\alpha{}_{,i} - \xi^i{}_{,\beta} u^\beta{}_{,n} u^\alpha{}_{,i}, \tag{16.40}$$

$$\begin{aligned}\eta^\alpha_{nm} = {} & \eta^\alpha{}_{,nm} + \eta^\alpha{}_{,n\beta} u^\beta{}_{,m} + \eta^\alpha{}_{,m\beta} u^\beta{}_{,n} - \xi^i{}_{,nm} u^\alpha{}_{,i} \\ & + \eta^\alpha{}_{,\beta\gamma} u^\beta{}_{,n} u^\gamma{}_{,m} - (\xi^k{}_{,n\beta} u^\beta{}_{,m} + \xi^k{}_{,m\beta} u^\beta{}_{,n}) u^\alpha{}_{,k} \\ & - \xi^k{}_{,\beta\gamma} u^\beta{}_{,n} u^\gamma{}_{,m} u^\alpha{}_{,k} + \eta^\alpha{}_{,\beta} u^\beta{}_{,nm} - \xi^i{}_{,n} u^\alpha{}_{,mi} - \xi^i{}_{,m} u^\alpha{}_{,ni} \\ & - \xi^k{}_{,\beta} (u^\alpha{}_{,k} u^\beta{}_{,nm} + u^\beta{}_{,n} u^\alpha{}_{,mk} + u^\alpha{}_{,nk} u^\beta{}_{,m}).\end{aligned} \tag{16.41}$$

This explicit expression for η^α_{nm} looks rather nasty and indicates that even the task of simply writing down the symmetry conditions $\mathbf{X}H_A = 0$ in detail can be a rather tiresome job. Imagine, for example, treating Einstein's vacuum field equations [see the later equations (16.67)–(16.69)]: 10 field equations ($A = 1, \ldots, 10$) for 10 functions ($\alpha, \beta, \ldots = 1, \ldots, 10$) of 4 variables ($n, m, k, \ldots = 1, \ldots, 4$) – how many terms will appear in the 10 symmetry conditions? So how can one proceed when a computer program for dealing with symmetries is not at hand?

The answer is that one should avoid ever writing down the symmetry conditions in full but instead try to extract the information piecewise, starting, for example, by taking into account *and writing down* only the terms with the highest derivatives occurring in the symmetry conditions

$$\mathbf{X}H_A \equiv 0 \qquad (\mathrm{mod}\, H_A = 0), \tag{16.42}$$

and then, with some information on ξ^n and η^α already gained, proceeding further with the derivatives of lower order. The simple differential equation

$$u_{,xx} + u_{,yy} = 0 \tag{16.43}$$

(the two-dimensional potential equation) will serve as an example of what we mean by this. Although this idea need not necessarily work for all differential equations, we shall put it into the form of an algorithm (which is a little bit pompous when applied to the potential equation).

1. Write the symmetry condition(s) without inserting the explicit expressions (16.40) and (16.41) for η^α_n and η^α_{nm}, respectively.

If we use the notation

$$x = x^1, \qquad y = x^2, \qquad H = u_{,11} + u_{,22} = 0,$$
$$\mathbf{X} = \xi^1 \frac{\partial}{\partial x^1} + \xi^2 \frac{\partial}{\partial x^2} + \eta \frac{\partial}{\partial u} + \cdots, \tag{16.44}$$

the symmetry condition reads

$$\mathbf{X}H \equiv \eta_{11} + \eta_{22} \equiv 0 \qquad (\mathrm{mod}\, u_{,11} + u_{,22} = 0). \tag{16.45}$$

2. Compute for the η^α_{nm} that actually appear in the symmetry condition(s) those terms that contain second derivatives $u^\alpha{}_{,nm}$ *not* occurring in the differential equation(s).

The second derivative $u_{,12}$ does not appear in $H = 0$, and the relevant terms of η_{11} and η_{22} are easily found to be

$$\begin{aligned} \eta_{11} &= -2\xi^2{}_{,1}u_{,12} - 2\xi^2{}_{,u}u_{,1}u_{,12} + \cdots, \\ \eta_{22} &= -2\xi^1{}_{,2}u_{,12} - 2\xi^1{}_{,u}u_{,2}u_{,12} + \cdots. \end{aligned} \tag{16.46}$$

3. Insert these terms into the symmetry conditions; since they cannot be eliminated or affected by using $H_A = 0$, they have to vanish separately.

From (16.45) and (16.46) we obtain

$$u_{,12}(\xi^2{}_{,1} + \xi^1{}_{,2}) + u_{,12}(u_{,1}\xi^2{}_{,u} + u_{,2}\xi^1{}_{,u}) = 0, \tag{16.47}$$

that is

$$\xi^1{}_{,u} = 0 = \xi^2{}_{,u}, \qquad \xi^2{}_{,1} + \xi^1{}_{,2} = 0. \tag{16.48}$$

4. Compute for the η^α_{nm} appearing in the symmetry conditions all terms containing second derivatives (using the above results), insert them into the symmetry conditions, and make them vanish by using $H_A = 0$, neglecting all terms in $H_A = 0$ that are not of second order.

Because of (16.48), most of the second order terms in the definition (16.41) of η_{nm} vanish. The surviving terms are

$$\begin{aligned} \eta_{11} &= -2\xi^1{}_{,1}u_{11} + \eta_{,u}u_{,11} + \cdots, \\ \eta_{22} &= -2\xi^2{}_{,2}u_{22} + \eta_{,u}u_{,22} + \cdots, \end{aligned} \tag{16.49}$$

and the symmetry condition $\eta_{11} + \eta_{22} = 0$, together with $u_{,11} + u_{,22} = 0$, yields

$$\xi^1{}_{,1} = \xi^2{}_{,2}. \tag{16.50}$$

5. Compute the η^α_{nm} appearing in the symmetry conditions completely, insert them into the symmetry conditions, and satisfy these conditions (taking $H_A = 0$ properly into account).

Taking into account the above results for ξ^1, ξ^2, and η, we have

$$\begin{aligned} \eta_{11} &= \eta_{,11} + 2\eta_{,1u}u_{,1} - \xi^1{}_{,11}u_{,1} - \xi^2{}_{,11}u_{,2} + \eta_{,uu}u_{,1}^2, \\ \eta_{22} &= \eta_{,22} + 2\eta_{,2u}u_{,2} - \xi^1{}_{,22}u_{,1} - \xi^2{}_{,22}u_{,2} + \eta_{,uu}u_{,2}^2, \end{aligned} \tag{16.51}$$

and the part of the symmetry conditions still to be satisfied is

$$\eta_{,11} + \eta_{,22} + 2u_{,1}\eta_{,1u} + 2u_{,2}\eta_{,2u} + (u_{,1}^2 + u_{,2}^2)\eta_{,uu} = 0, \tag{16.52}$$

since we know already from (16.48) and (16.50) that

$$\xi^1{}_{,11} + \xi^1{}_{,22} = 0 = \xi^2{}_{,11} + \xi^2{}_{,22}. \tag{16.53}$$

Equation (16.52) has to be satisfied identically in the derivatives $u_{,1}$ and $u_{,2}$ and thus leads to

$$\eta = a_1 u + \mathring{\eta}, \qquad \mathring{\eta}_{,11} + \mathring{\eta}_{,22} = 0. \tag{16.54}$$

Putting all the pieces together, we find that the symmetry generator of the two-dimensional potential equation has the form

$$\mathbf{X} = \xi^1(x^1, x^2)\frac{\partial}{\partial x^1} + \xi^2(x^1, x^2)\frac{\partial}{\partial x^2} + \mathring{\eta}(x^1, x^2)\frac{\partial}{\partial u} + a_1 u\frac{\partial}{\partial u} \tag{16.55}$$

with

$$\xi^1{}_{,2} = -\xi^2{}_{,1}, \qquad \xi^1{}_{,1} = \xi^2{}_{,2}, \qquad \mathring{\eta}_{,11} + \mathring{\eta}_{,22} = 0. \tag{16.56}$$

Obviously, the first part of (16.56) are the Cauchy–Riemann differential equations, and ξ^1 and ξ^2 are the real and the imaginary part of an analytic function, respectively. To simplify (16.55) and (16.56), one can therefore introduce new variables ζ, B and their complex conjugates $\bar{\zeta}$, $\bar{B}$ by

$$\zeta = x + iy, \qquad B = \xi^1 + i\xi^2, \qquad \mathring{\eta} = W(\zeta) + \bar{W}(\bar{\zeta}). \tag{16.57}$$

In these variables, the potential equation reads

$$u_{,\zeta\bar{\zeta}} = 0 \tag{16.58}$$

and its symmetries are given by

$$\mathbf{X} = B(\zeta)\frac{\partial}{\partial \zeta} + \bar{B}(\bar{\zeta})\frac{\partial}{\partial \bar{\zeta}} + [W(\zeta) + \bar{W}(\bar{\zeta})]\frac{\partial}{\partial u} + a^1 u\frac{\partial}{\partial u} \tag{16.59}$$

with arbitrary analytic functions B and W. B corresponds to an infinitesimal conformal transformation $\tilde{\zeta} = f(\zeta)$, W to the addition of a solution of the (linear!) potential equation, and the constant a_1 to the multiplication of a solution by a constant factor.

If we apply the above routine to the *heat conduction equation*

$$u_{,11} - u_{,2} = 0 \qquad (x^1 \equiv x, x^2 \equiv t) \tag{16.60}$$

and the symmetry generator

$$\mathbf{X} = \xi^1\frac{\partial}{\partial x^1} + \xi^2\frac{\partial}{\partial x^2} + \eta\frac{\partial}{\partial u} + \cdots \tag{16.61}$$

it might admit, we obtain from steps 1–4

$$\xi^2{}_{,1} = 0, \qquad \xi^2{}_{,u} = 0, \tag{16.62}$$

and the remaining terms of the symmetry conditions yield

$$\begin{aligned} &\xi^1{}_{,u} = 0, \qquad \eta_{,uu} = 0, \qquad 2\xi^1{}_{,1} = \xi^2{}_{,2}, \\ &2\eta_{,1u} - \xi^1{}_{,11} + \xi^1{}_{,2} = 0, \qquad \eta_{,11} = \eta_{,2}. \end{aligned} \tag{16.63}$$

The system (16.62) and (16.63) can easily be solved. We leave the details to the reader and give only the result, which is

$$\begin{aligned} \mathbf{X} &= (a_1 + a_2 x + a_3 t + a_4 tx)\frac{\partial}{\partial x} + (2a_2 t + a_4 t^2 + a_5)\frac{\partial}{\partial t} \\ &\quad + [-\tfrac{1}{2}a_3 xu - a_4(\tfrac{1}{2}ut + \tfrac{1}{4}ux^2) + a_6 u + g(x,t)]\frac{\partial}{\partial u}, \end{aligned} \tag{16.64}$$

where $a_1, \ldots, a_6$ are arbitrary constants, and $g(x, t)$ is a solution of the heat conduction equation. To understand the physical (or mathematical) meaning of the different parts of the generator (16.64), one could, for example, construct and study the finite transformations correspondng to these parts. We shall come back to this point in Section 17.3 and will not pursue this idea here. [By the way, the terms with a_3 correspond to a Galilei transformation, and those with a_4 to a projective transformation in (x, t)-space.]

At the end of this section we want to give some more examples of partial differential equations, simply in the form of a list, without going into the details of the derivation from the symmetry conditions.

The *sine–Gordon equation*

$$u_{,xx} - u_{,tt} = \sin u \tag{16.65}$$

admits the generator

$$\mathbf{X} = (a_1 + a_2 t)\frac{\partial}{\partial x} + (a_2 x + a_3)\frac{\partial}{\partial t} \tag{16.66}$$

of the translations and Lorentz rotations in the x–t plane.

Einstein's field equations for the vacuum,

$$R_{ab} = 0, \qquad a, b = 1, \ldots, 4, \tag{16.67}$$

with

$$R_{ab} = R^n{}_{anb},$$
$$R^n{}_{amb} = \Gamma^n_{ab,m} - \Gamma^n_{am,b} + \Gamma^n_{rm}\Gamma^r_{ab} - \Gamma^n_{rb}\Gamma^r_{am}, \tag{16.68}$$
$$\Gamma^n_{ab} = \tfrac{1}{2}g^{nc}(g_{ac,b} + g_{bc,a} - g_{ab,c}), \qquad g^{nc}g_{cm} = \delta^n_m,$$

are a system of 10 second order differential equations for the 10 metric functions $g_{ab}(x^n)$, which are functions of 4 variables x^n. The generator of the general point symmetry of this system reads

$$\mathbf{X} = \xi^i(x^n)\frac{\partial}{\partial x^i} - (\xi^k{}_{,i}g_{kn} + \xi^k{}_{,n}g_{ik} + ag_{in})\frac{\partial}{\partial g_{in}}, \tag{16.69}$$

where the four arbitrary functions $\xi^i(x^n)$ correspond to coordinate transformations $\tilde{x}^n = \tilde{x}^n(x^i)$ and the constant a to a multiplication of the functions g_{ab} by a constant.

For a special subcase (the Robinson–Trautman solutions of Petrov type III), these field equations reduce to

$$P(P_{,xx} + P_{,yy}) - (P^2_{,x} + P^2_{,y}) + 3x\sqrt{2} = 0, \tag{16.70}$$

which admits

$$\mathbf{X} = (ax + bxy)\frac{\partial}{\partial x} + [ay + \tfrac{1}{2}b(y^2 - x^2) + c]\frac{\partial}{\partial y}$$
$$+ (a\tfrac{3}{2}P + b\tfrac{3}{2}yP)\frac{\partial}{\partial P} \tag{16.71}$$

as point symmetries.

We end with an example of a third order differential equation, the *Korteweg–deVries equation*

$$u_{,xxx} + 6uu_{,x} + u_{,t} = 0. \tag{16.72}$$

Its symmetry generators are

$$\mathbf{X} = (a_1 + a_2x + 6a_3t)\frac{\partial}{\partial x} + (3a_2t + a_4)\frac{\partial}{\partial t} + (a_3 - 2a_2u)\frac{\partial}{\partial u}, \tag{16.73}$$

corresponding to translations (a_1, a_4), to a scaling (a_2), and to a Galilei transformation (a_3).

The mere listing of so many differential equations and their symmetries may have left the reader a little bit bewildered, and those who try to verify, for example, the symmetries of Einstein's equation may also feel exhausted by the complexity of the computations (compared with the apparently meagre result). But in the next chapters (when dealing with applications), we hope to convince the reader that it pays to look for point symmetries when trying to solve partial differential equations.

16.3 Exercises

1 Find the symmetries of the first order equation $b^0(x^i) + b^n(x^i)\varphi_{,n} = 0$.

2 Find the symmetries of the first order system $u_{,x} = v_{,y}$, $u_{,y} = -v_{,x}$, write them in terms of $w = v + iu$, $\zeta = x + iy$ and their complex conjugates, and compare the result with (16.59) valid for the potential equation (16.58).

3 Check that (16.71) are all the symmetries of the differential equation (16.70).

17

How to use Lie point symmetries of partial differential equations I: generating solutions by symmetry transformations

17.1 The structure of the set of symmetry generators

As the examples of the preceding chapter show, the symmetry generators of partial differential equations clearly divide into two classes: they may depend either on arbitrary constants or on arbitrary functions. The two will in general occur simultaneously, as the examples of the heat conduction equation

$$\mathbf{X} = (a_1 + a_2 x + a_3 t + a_4 t x)\frac{\partial}{\partial x} + (2a_2 t + a_4 t^2 + a_5)\frac{\partial}{\partial t}$$

$$+ [-\tfrac{1}{2}a_3 x u - a_4(\tfrac{1}{2}ut + \tfrac{1}{4}ux^2) + a_6 u + g(x,t)]\frac{\partial}{\partial u}, \qquad (17.1)$$

$$g_{,xx} - g_{,t} = 0$$

and Einstein's equations

$$\mathbf{X} = \xi^i(x^n)\frac{\partial}{\partial x^i} - (\xi^k{}_{,i} g_{kn} + \xi^k{}_{,n} g_{ik} + a_1 g_{in})\frac{\partial}{\partial g_{in}} \qquad (17.2)$$

show [although the function $g(x,t)$ in (17.1) is subject to the heat conduction equation, its initial values $g(x,0)$ are arbitrary].

That part of a symmetry generator that depends on arbitrary constants will give rise to a Lie algebra (as was the case for Lie point symmetries of ordinary differential equations): if there are r independent parameters a_N, then the part in question can be written as

$$\mathbf{X} = a_N \mathbf{X}_N, \qquad N = 1, \ldots, r, \tag{17.3}$$

and the commutators of the $\mathbf{X}_N$ obey

$$[\mathbf{X}_N, \mathbf{X}_M] = C^A_{NM} \mathbf{X}_A \tag{17.4}$$

with constant coefficients (structure constants) C^A_{NM}. So the generator (17.1) leads to a six-parameter Lie algebra with the basis

$$\begin{aligned} &\mathbf{X}_1 = \frac{\partial}{\partial x}, && \mathbf{X}_4 = tx\frac{\partial}{\partial x} + t^2\frac{\partial}{\partial t} - (\tfrac{1}{2}ut + \tfrac{1}{4}ux^2)\frac{\partial}{\partial u}, \\ &\mathbf{X}_2 = x\frac{\partial}{\partial x} + 2t\frac{\partial}{\partial t}, && \mathbf{X}_5 = \frac{\partial}{\partial t}, \\ &\mathbf{X}_3 = t\frac{\partial}{\partial x} - \tfrac{1}{2}xu\frac{\partial}{\partial u}, && \mathbf{X}_6 = u\frac{\partial}{\partial u}, \end{aligned} \tag{17.5}$$

whereas (17.2) contains only the one-parameter algebra with basis

$$\mathbf{X} = g_{in}\frac{\partial}{\partial g_{in}}. \tag{17.6}$$

When there are arbitrary functions as elements of a symmetry generator, then the set of *all* symmetries does not form a Lie algebra; that is, it is not possible to represent the symmetry as $\mathbf{X} = a_N \mathbf{X}_N$ with constant a_N and a *finite* number of basic symmetries $\mathbf{X}_N$ that obey a commutator relation (17.4) with constant C^A_{NM}. But in that case too the commutator of two symmetry generators is again a generator of a symmetry (we leave the proof to the reader). It may even be possible to construct a finite dimensional Lie algebra by assigning special values to the arbitrary functions appearing in the generators. In the case of the generator (17.2) this can, for example, be done by choosing $\xi^i = \delta^i_N$, that is,

$$\mathbf{X}_N = \delta^i_N \frac{\partial}{\partial x^i}, \qquad i, N = 1, \ldots, 4. \tag{17.7}$$

But note that these $\mathbf{X}_N$ *and* the $\mathbf{X}$ of (17.6) do *not* form a Lie algebra!

As the two examples (17.1) and (17.2) given above illustrate, the occurrence of arbitrary functions in the symmetry generators can have various causes. For a linear differential equation such as the heat conduction equation, the arbitrary function enters because the addition of an arbitrary solution to a given solution will again give a solution. For Einstein's field equation, the arbitrary functions $\xi^i(x^n)$ are due to the gauge freedom inherent in Einstein's theory: the theory is invariant (covariant) with respect to arbitrary coordinate transformations; such transformations may change the form but not the physical meaning of a solution. Therefore, those transformations are considered trivial, although mathematically they lead to new solutions.

17.2 What can symmetry transformations be expected to achieve?

For ordinary differential equations, the existence of Lie point symmetries always ensured the possibility of going a step forward in the integration of that equation, and if there was a sufficiently large number of symmetries, the complete integration of the differential equation was guaranteed in most cases.

For partial differential equations the situation is different, and we want to make a few remarks to explain why.

The general solution of an nth order ordinary differential equation depends on n arbitrary constants. If that differential equation admits an r-parameter group of Lie point symmetries, a general symmetry transformation (and its generator) will depend on r arbitrary parameters. So, loosely speaking, the application of a symmetry transformation will provide us with r of the n necessary constants of integration, and for $r<n$ we can reduce the order of the differential equation or for $r \geqslant n$ even solve it completely.

The general solution of a partial differential equation will usually depend on one or several arbitrary *functions*. Symmetry generators depending on a finite number of arbitrary *constants* may lead to solutions involving arbitrary constants, but there is no hope of obtaining some (or the) general solution or of lowering the order of the differential equation by those symmetries. Symmetry generators depending on arbitrary *functions* seem to offer better prospects, but in practice they also give no easy way to obtain a general solution in closed form: for a linear partial differential equation the task of finding the general symmetry is of equal difficulty to solving the original differential equation (the arbitrary function appearing in the symmetry generator is just a solution of that equation!), and although arbitrary functions in the generators that correspond to gauge transformations lead to arbitrary functions in the solutions, these degrees of freedom are exactly those that are of no interest in applications.

So we can only hope to obtain some special finite-parameter classes of solutions of a partial differential equation when applying its Lie point symmetry transformations to already known solutions. The details of this idea will be discussed immediately. A completely different idea for using Lie point symmetries will be treated in the following chapters.

17.3 Generating solutions by finite symmetry transformations

By definition, a symmetry transformation maps a solution into a solution. So if we already know a (special) solution of a partial differential equation, we can apply a finite symmetry transformation to obtain a (possibly) new solution. This new solution will depend on at most as many new parameters as there are in the symmetry transformation we have used.

To carry out this idea, we need the finite transformations

$$\tilde{x}^n = \tilde{x}^n(x^i, u^\beta; \varepsilon), \qquad \tilde{u}^\alpha = \tilde{u}^\alpha(x^i, u^\beta; \varepsilon), \tag{17.8}$$

corresponding to the infinitesimal generator

$$\mathbf{X} = \xi^n(x^i, u^\beta)\frac{\partial}{\partial x^n} + \eta^\alpha(x^i, u^\beta)\frac{\partial}{\partial u^\alpha}. \tag{17.9}$$

Geometrically speaking, we need a parametric representation of those lines in (x^n, u^α)-space whose tangent vectors at each point are given by the components (ξ^n, η^α) of $\mathbf{X}$. That is, these lines must satisfy the first order differential equations

$$\frac{\mathrm{d}\tilde{x}^n}{\mathrm{d}\varepsilon} = \xi^n[\tilde{x}^i(\varepsilon), \tilde{u}^\beta(\varepsilon)], \qquad \frac{\mathrm{d}\tilde{u}^\alpha}{\mathrm{d}\varepsilon} = \eta^\alpha[\tilde{x}^i(\varepsilon), \tilde{u}^\beta(\varepsilon)] \tag{17.10}$$

with initial values

$$\tilde{x}^n = x^n, \qquad \tilde{u}^\alpha = u^\alpha \quad \text{for } \varepsilon = 0. \tag{17.11}$$

Here ε is the parameter along the lines; its value indicates how far we have gone in (x^i, u^β)-space in the direction determined by the vector $\mathbf{X}$. Like the generator $\mathbf{X}$, so this direction can depend on arbitrary constants and/or functions, which will then appear again in $\tilde{x}^n$ and $\tilde{u}^\alpha$. If we know the finite transformations (17.8), we also have to hand the inverse transformations

$$x^i = x^i(\tilde{x}^n, \tilde{u}^\alpha; \varepsilon), \qquad u^\beta = u^\beta(\tilde{x}^n, \tilde{u}^\alpha; \varepsilon). \tag{17.12}$$

To apply the symmetry transformation to a special solution

$$u^\beta = f^\beta(x^i) \Leftrightarrow F^\delta(u^\beta, x^i) = 0, \tag{17.13}$$

one simply has to replace x^i and u^β by means of (17.12) to obtain

$$F^\delta[u^\beta(\tilde{x}^n, \tilde{u}^\alpha; \varepsilon), x^i(\tilde{x}^n, \tilde{u}^\alpha; \varepsilon)] = 0, \tag{17.14}$$

which can be resolved with respect to the $\tilde{u}^\alpha$ to yield the transformed and hopefully new solution

$$\tilde{u}^\alpha = \tilde{u}^\alpha(\tilde{x}^n; \varepsilon). \tag{17.15}$$

We now want to illustrate this general procedure by a few examples. The main difficulty is of course the integration of the system (17.10) of first order differential equations, but in many applications the components ξ^n, η^α will be simple functions of x^i, u^β, and the integration will not be a serious problem.

We start with the heat conduction equation $u_{,xx} - u_{,t} = 0$ and its symmetry generator

$$\mathbf{X} = xt\frac{\partial}{\partial x} + t^2\frac{\partial}{\partial t} - \left(\frac{x^2}{4} + \frac{t}{2}\right)u\frac{\partial}{\partial u}; \tag{17.16}$$

compare Section 16.2. For this special symmetry, the differential equations (17.10) for the finite transformations $\tilde{x}$, $\tilde{t}$, and $\tilde{u}$ read

$$\frac{\mathrm{d}\tilde{x}}{\mathrm{d}\varepsilon} = \tilde{x}\tilde{t}, \qquad \frac{\mathrm{d}\tilde{t}}{\mathrm{d}\varepsilon} = \tilde{t}^2, \qquad \frac{\mathrm{d}\tilde{u}}{\mathrm{d}\varepsilon} = -\tilde{u}\left(\frac{\tilde{x}^2}{4} + \frac{\tilde{t}}{2}\right). \tag{17.17}$$

The first two of these equations yield $\mathrm{d}\tilde{x}/\tilde{x} = \mathrm{d}\tilde{t}/\tilde{t}$, which is integrated by

$$\tilde{x}/x = \tilde{t}/t \tag{17.18}$$

(remember that x, t, and u are the initial values of $\tilde{x}$, $\tilde{t}$ and $\tilde{u}$, respectively). Substituting this back into (17.17), we obtain $\mathrm{d}\tilde{x}/\mathrm{d}\varepsilon = \tilde{x}^2 t/x$ and $\mathrm{d}\tilde{t}/\mathrm{d}\varepsilon = \tilde{t}^2$ and thus

$$\tilde{x} = x/(1 - \varepsilon t), \qquad \tilde{t} = t/(1 - \varepsilon t). \tag{17.19}$$

The last of the three equations (17.17) can now be written as

$$\frac{\mathrm{d}\tilde{u}}{\tilde{u}} = -\frac{x^2\,\mathrm{d}\varepsilon}{4(1-\varepsilon t)^2} - \frac{t\,\mathrm{d}\varepsilon}{2(1-\varepsilon t)}, \tag{17.20}$$

which yields

$$\tilde{u} = u\sqrt{1 - \varepsilon t}\,e^{-\varepsilon x^2/4(1-\varepsilon t)}. \tag{17.21}$$

Equations (17.19) and (17.20) are the finite symmetry transformations.

Their inverse is

$$x = \tilde{x}/(1+\varepsilon\tilde{t}), \qquad t = \tilde{t}/(1+\varepsilon\tilde{t}), \qquad u = \tilde{u}\sqrt{1+\varepsilon\tilde{t}}\,e^{\varepsilon\tilde{x}^2/4(1+\varepsilon\tilde{t})}, \tag{17.22}$$

simply obtained by exchanging (x, t, u) and $(\tilde{x}, \tilde{t}, \tilde{u})$ and replacing ε by $-\varepsilon$.

Now we can apply this symmetry transformation to any solution of the heat conduction equation, for example, to the trivial solution $u = A = \text{const}$. We immediately get, from this and (17.22),

$$u = \frac{A}{\sqrt{1+\varepsilon t}}\,e^{-\varepsilon x^2/4(1+\varepsilon t)} \tag{17.23}$$

(after having dropped the tilde); that is, we have generated a nontrivial two-parameter solution from the one-parameter solution $u = A$.

The symmetry generator

$$\mathbf{X} = (a_1 + a_2 t)\frac{\partial}{\partial x} + a_3\frac{\partial}{\partial t} - \frac{a_2}{2}xu\frac{\partial}{\partial u} \tag{17.24}$$

of the heat conduction equation can be treated in a similar way. The finite transformations corresponding to it are easily obtained from

$$\frac{d\tilde{x}}{d\varepsilon} = a_1 + a_2\tilde{t}, \qquad \frac{d\tilde{t}}{d\varepsilon} = a_3, \qquad \frac{d\tilde{u}}{d\varepsilon} = -\frac{a_2}{2}\tilde{x}\tilde{u} \tag{17.25}$$

(starting with the second of these three equations). They read

$$\begin{aligned} \tilde{x} &= x + a_2\varepsilon t + a_1\varepsilon + \tfrac{1}{2}a_2 a_3\varepsilon^2, \\ \tilde{t} &= t + a_3\varepsilon, \\ \tilde{u} &= u\exp\left(-\tfrac{1}{2}a_2\varepsilon x - \tfrac{1}{4}a_2^2\varepsilon^2 t - \tfrac{1}{4}a_1 a_2\varepsilon^2 - \tfrac{1}{12}a_2^2 a_3\varepsilon^3\right). \end{aligned} \tag{17.26}$$

If we apply this symmetry transformation again to the trivial solution $u = A = \text{const.}$, we obtain (having dropped the tilde)

$$\begin{aligned} &u = \hat{A}\exp(Cx + C^2 t), \qquad \hat{A} = A\exp\left(\tfrac{1}{4}a_1 a_2\varepsilon^2 + \tfrac{1}{12}a_2^2 a_3\varepsilon^3\right), \\ &C = \tfrac{1}{2}a_2\varepsilon. \end{aligned} \tag{17.27}$$

Although the symmetry transformation depends on three parameters a_1, a_2, and a_3, the new solution contains only the two independent parameters $\hat{A}$ and C; only one new parameter is generated. We can easily trace back the reason for this lack of more parameters: if we perform the finite transformation with $a_1 = 0 = a_3$, we arrive at

$$u = A \exp\left(\tfrac{1}{2}a_2\varepsilon x + \tfrac{1}{4}a_2^2\varepsilon^2 t\right). \tag{17.28}$$

The generators $a_1\,\partial/\partial x$ and $a_3\,\partial/\partial t$ effect translations in the x- and t-directions, respectively, but adding constants to x and t in (17.28) is equivalent to changing the parameter A. For less trivial solutions of the heat conduction equation, the finite symmetries discussed above may lead to more complicated solutions.

The third and last example concerns the differential equation $P(P_{,xx} + P_{,yy}) - (P^2{}_{,x} + P^2{}_{,y}) + 3x\sqrt{2} = 0$ and its most general Lie point symmetry (16.71). The corresponding first order differential equations are

$$\begin{aligned} \frac{\mathrm{d}\tilde{x}}{\mathrm{d}\varepsilon} &= (a_1 + a_2\tilde{y})\tilde{x}, \\ \frac{\mathrm{d}\tilde{y}}{\mathrm{d}\varepsilon} &= a_1\tilde{y} + \tfrac{1}{2}a_2(\tilde{y}^2 - \tilde{x}^2) + a_3, \\ \frac{\mathrm{d}\tilde{P}}{\mathrm{d}\varepsilon} &= \tfrac{3}{2}\tilde{P}(a_1 + a_2\tilde{y}). \end{aligned} \tag{17.29}$$

Although this system can be integrated, we shall not try to go ahead with the integration here. The reason is that the only known solution to the partial differential equation in question is

$$P = 2^{3/4}x^{3/2}, \tag{17.30}$$

and since an immediate consequence of (17.29) is

$$\frac{\mathrm{d}}{\mathrm{d}\varepsilon}(\tilde{P}\tilde{x}^{-3/2}) = 0, \tag{17.31}$$

it is impossible to change the solution (17.30) by means of any finite symmetry transformation; a new solution cannot be generated. The function $w \equiv Px^{-3/2}$ is an invariant of the group generated by the three generators (16.71); compare Sections 6.4 and 19.1.

17.4 Generating solutions (of linear differential equations) by applying the generators

As already mentioned and discussed, for linear differential equations the general symmetry generator **X** contains functions that are arbitrary solutions of that equation, corresponding to the possibility of adding an arbitrary solution to a given one to obtain a new solution. The generators then have the typical structure

$$\mathbf{X} = a_A \mathbf{X}_A + f^\alpha(x^i)\frac{\partial}{\partial u^\alpha}, \tag{17.32}$$

where f^α is a solution of the differential equation, a_A are arbitrary constants, and the $\mathbf{X}_A$ form a Lie algebra.

Although generators of the type (17.32) do not form a Lie algebra, a commutator of two such symmetries is again a symmetry; compare also Exercise 17.1. So with **X** given by (17.32) and **Y** given as

$$\mathbf{Y} = b_A \mathbf{X}_A + g^\alpha(x^i)\frac{\partial}{\partial u^\alpha}, \tag{17.33}$$

the commutator of these two necessarily has the same form

$$[\mathbf{X}, \mathbf{Y}] = c_A \mathbf{X}_A + h^\alpha(x^i)\frac{\partial}{\partial u^\alpha}. \tag{17.34}$$

Evaluating the left side of this relation, we see that the functions h^α can be expressed in terms of f^α, g^α, the components of $\mathbf{X}_A$, and the derivatives of these functions. But the h^α must be again solutions, and so the relation (17.34) offers a way of constructing a new solution h^α from one (e.g., for $f^\alpha = 0$) or two known solutions merely by the process of differentiation.

To see how this works in practice, let us take the heat conduction equation $u_{,xx} = u_{,t}$ and consider the commutators (17.34) for the case $\mathbf{X} = \mathbf{X}_A$, $\mathbf{Y} = g\,\partial/\partial u$ (i.e., $b_A = 0 = f^\alpha$), with $\mathbf{X}_A$ given by (17.5) and, of course, g a solution ($g_{,xx} = g_{,t}$). It is easy to extract from

$$\left[\mathbf{X}_A, g(x,t)\frac{\partial}{\partial u}\right] = c_{AB}\mathbf{X}_B + h_A(x,t)\frac{\partial}{\partial u} \tag{17.35}$$

the different functions $h_A(x, t)$ (all the c_{AB} vanish). The result is

$$\begin{aligned} h_1 &= g_{,x}, & h_4 &= txg_{,x} + t^2 g_{,t} + (\tfrac{1}{2}t + \tfrac{1}{4}x^2)g, \\ h_2 &= xg_{,x} + 2tg_{,t}, & h_5 &= g_{,t}, \\ h_3 &= tg_{,x} + \tfrac{1}{2}xg, & h_6 &= -g. \end{aligned} \tag{17.36}$$

All these functions h_A are solutions to the heat conduction equation, provided $g(x, t)$ is a solution. This procedure can of course be iterated, starting with one of the h_i instead of g, thus leading to an infinite number of solutions.

For a linear differential equation *with constant coefficients* – like the heat conduction equation – the resulting new solutions are often rather trivially related to the original input solutions, but for more complicated cases the results can be more interesting.

17.5 Exercises

1 Show that if $\mathbf{X}$ and $\mathbf{Y}$ are symmetry generators, then the commutator $[\mathbf{X}, \mathbf{Y}]$ is too.

2 Is it possible to specify the function $g(x, t)$ in (17.1) so that $\mathbf{X}_7 = g(x, t)\,\partial/\partial u$ together with the six generators (17.5) is the Lie algebra of a seven-parameter Lie group?

3 Find the finite transformation for $\mathbf{X} = (a_1 + 6t)\,\partial/\partial x + a_4\,\partial/\partial t + \partial/\partial u$ and apply it to the special solution $u = -2/x^2$ of the Korteweg–deVries equation (16.72).

18

How to use Lie point symmetries of partial differential equations II: similarity variables and reduction of the number of variables

18.1 The problem

Solutions of partial differential equations depend on arbitrary functions, represented, for example, by initial or boundary values, and since reality is (3 + 1)-dimensional (three spacelike and one timelike dimension), these functions will be functions of three variables if general solutions of physically important differential equations are being treated.

If we want to give these solutions in terms of "known" or even "simple" functions, or if we just want to find exact solutions, we practically always aim to express the solution in terms of functions of *one* (sometimes complex) variable. There are two main streams of dealing with partial differential equations and their solutions which are based on this experience.

The first is to try a separation *ansatz*, that is, to look for solutions that are products (or sums) of functions of one variable (and perhaps to give the general solutions in terms of such products). We shall come back to this problem in Chapter 20.

The second is to ask for solutions that depend on less variables than occur in the original formulation of the differential equation, maybe even on one variable alone. Typical examples are spherically symmetric solutions to the potential equation, or plane-fronted monochromatic waves in Maxwell theory. Obviously, the problem we have to solve is the following. Given a set of differential equations

$$H_A(x^n, u^\alpha, u^\alpha{}_{,n}, \ldots) = 0, \qquad n = 1, \ldots, N, \quad \alpha = 1, \ldots, M, \tag{18.1}$$

in the $N+M$ variables $\{x^n, u^\alpha\}$, is it possible to find solutions that involve only $N+M-1$ or even less variables? We have deliberately put the independent and dependent variables on an equal footing since it may be necessary first to make a variable transformation before, say, looking for solutions independent of the (new) variable x^1.

Simply to take the system (18.1) and ask for solutions that are, for example, independent of x^1 may lead to contradictions as the example

$$x^2(u_{,y}-1)-u^2{}_{,x}+u^2=0, \qquad u=u(x,y), \tag{18.2}$$

shows. If we assume $u_{,x}=0$, then (18.2) reduces to

$$x^2(u_{,y}-1)+u^2=0, \tag{18.3}$$

which does not admit solutions independent of x. Are there methods to ensure from the very beginning that assuming the solution is independent of a particular variable does not lead to such contradictions, and are there methods to find such variables? As we shall see now, the symmetry approach provides the necessary tools to answer these questions.

18.2 Similarity variables and how to find them

The obvious reason that the assumption $u_{,x}=0$ led to a contradiction to the differential equation (18.2) is that the differential equation depends explicitly on x, whereas its solution was assumed to be independent of that variable. If the original equation and its consequence (18.3) were independent of x, no such difficulty could have occurred. That is, if $\mathbf{X}=\partial/\partial x$ is a symmetry generator of a differential equation, the assumption $u_{,x}=0$ cannot lead to an inconsistency since from $\mathbf{X}H=0 \pmod{H=0}$ it follows that the differential equation can be written in a form such that H does not depend on x; compare the remarks following equation (15.31).

Since any symmetry generator $\mathbf{X}$ can be transformed to its normal form

$$\mathbf{X}=\frac{\partial}{\partial\psi} \tag{18.4}$$

by a suitable transformation of variables, the above observation offers a straightforward way to find those variables (there may be several if there are several symmetries) from which the solutions can be assumed to be independent. One first has to introduce the $N+M$ new variables $\psi(x^i, u^\beta)$ and $\varphi^\Omega(x^i, u^\beta)$ defined by

$$\mathbf{X}\psi=1, \qquad \mathbf{X}\varphi^\Omega=0, \qquad \Omega=1,\ldots,N+M-1, \tag{18.5}$$

instead of the original N independent and M dependent variables x^i and u^β; compare also Section 2.2. From that set of new variables one should choose an appropriate set of N variables $\{\psi, y^n\}$ as independent and the rest of the φ^Ω, called w^α, as dependent variables. It will then be possible to rewrite the original system $H_A = 0$ of differential equations as

$$G_A(y^n, w^\alpha, \partial w^\alpha/\partial y^n, \partial w^\alpha/\partial\psi, \ldots) = 0, \tag{18.6}$$

where the G_A do *not* depend on ψ, so that we can assume $\partial w^\alpha/\partial\psi = 0$, that is, $w^\alpha = w^\alpha(y^n)$, without running into difficulties.

So we see that the existence of a symmetry ensures the possibility of diminishing by one the number of variables on which the solution depends. The adapted variables $\{\varphi^\Omega\} = \{w^\alpha, y^n\}$ are called *similarity variables* since the scaling invariance, that is, the invariance under similarity transformations, was one of the first examples where these variables were treated and used systematically. Correspondingly, the solutions $w^\alpha = w^\alpha(y^n)$, which are independent of ψ and thus invariant under the action of $\mathbf{X}$, are called *similarity solutions.* Of course, these similarity solutions are invariant under $\mathbf{X}$ in the original variables $\{x^i, u^\alpha\}$ too; that is, the surface $0 = u^\alpha - u^\alpha(x^n)$ in (x^i, u^α)-space will remain unchanged, and

$$\mathbf{X}[u^\alpha - u^\alpha(x^n)] = \eta^\alpha - \xi^n u^\alpha{}_{,n} = 0 \tag{18.7}$$

will hold for these solutions.

A difficulty in introducing similarity variables can arise when the variable ψ necessarily – because of, for example, its physical interpretation – has to be taken as a dependent variable, as is the case for the heat conduction equation and the symmetry generator $\mathbf{X} = \partial/\partial u$. It does not then make sense to ask for solutions independent of u, and this particular symmetry cannot be used to reduce the number of independent variables.

Looking back at the program sketched so far, it is clear that the main difficulty is in finding ψ and the similarity variables φ^Ω, that is, in solving $\mathbf{X}\psi = 1$ and the linear partial differential equation

$$\mathbf{X}\varphi = \left[\xi^n(x^i, u^\beta)\frac{\partial}{\partial x^n} + \eta^\alpha(x^i, u^\beta)\frac{\partial}{\partial u^\alpha}\right]\varphi(x^i, u^\beta) = 0. \tag{18.8}$$

Because any function $\phi(\varphi^\Omega)$ is again a similarity variable, similarity variables are not uniquely defined: what we need is a set of $N + M - 1$ functionally independent solutions φ^Ω to $\mathbf{X}\varphi = 0$. There are two main approaches to finding this set of solutions.

The first is to directly attack the differential equation (18.8), often by some intelligent guesswork. Due to the simple structure of many of the generators

X, this way is more promising than it may seem at first glance.

The second, more systematic way is to make use of the finite symmetry transformations

$$\tilde{x}^n = \tilde{x}^n(x^i, u^\beta, \varepsilon), \qquad \tilde{u}^\alpha = \tilde{u}^\alpha(x^i, u^\beta, \varepsilon), \tag{18.9}$$

which can be obtained by solving

$$\frac{d\tilde{x}^n}{d\varepsilon} = \xi^n(\tilde{x}^i, \tilde{u}^\beta), \qquad \frac{d\tilde{u}^\alpha}{d\varepsilon} = \eta^\alpha(\tilde{x}^i, \tilde{u}^\beta) \tag{18.10}$$

with initial values $\tilde{u}^\alpha = u^\alpha$, $\tilde{x}^n = x^n$ for, say, $\varepsilon = 0$; compare also Sections 2.2 and 17.3. For a fixed set of initial values, these finite transformations give an analytic expression for the curve that connects all points in $(\tilde{x}^n, \tilde{u}^\alpha)$-space, which can be transformed into these initial points by the action of the one-parameter group generated by **X**, ε being the parameter along this curve. For varying initial values, equations (18.9) give the set of all such curves. Of course, these curves are nothing other than the orbits of the group, and (18.9) gives a parametric representation of the orbits. But the solutions φ^Ω of $\mathbf{X}\varphi = 0$ exactly define those orbits; compare Section 6.4! So if we know the finite transformations (18.9), it is easy to extract the functions φ^Ω from them: we only have to eliminate the parameter ε from the parametric representation (18.9).

To do this, we take one of the equations (18.9), say, for $\xi^1 \neq 0$, $\tilde{x}^1 = \tilde{x}^1(x^i, u^\beta; \varepsilon)$, and solve it for ε,

$$\varepsilon = \varepsilon(\tilde{x}^1, x^i, u^\beta). \tag{18.11}$$

We then substitute this expression for ε back into equations (18.9) and obtain the implicit representation of the orbits

$$\begin{aligned} \tilde{x}^n &= \tilde{x}^n(x^i, u^\beta; \varepsilon(\tilde{x}^1, x^i, u^\beta)) = \tilde{x}^n(\tilde{x}^1, \varphi^\Omega), \\ \tilde{u}^\alpha &= \tilde{u}^\alpha(x^i, u^\beta; \varepsilon(\tilde{x}^1, x^i, u^\beta)) = \tilde{u}^\alpha(\tilde{x}^1, \varphi^\Omega). \end{aligned} \tag{18.12}$$

From these equations a set of similarity variables φ^Ω can in principle be read off. The ambiguity in doing this, that is, in choosing the functions $\varphi^\Omega(x^i, u^\beta)$, arises naturally since similarity variables are not uniquely defined.

What is still missing is the function $\psi(x^i, u^\beta)$ defined by $\mathbf{X}\psi = 1$. ψ is also not uniquely defined: we can add to a given ψ any function of the φ^Ω. Moreover, since our main goal is to construct solutions independent of ψ, it does not really matter for the intermediate steps whether we use ψ or any function of it. Because $\mathbf{X} = \partial/\partial\psi$, ψ varies along each orbit and parametrizes the points on an orbit as ε does: with $\varepsilon = \psi$, equation (18.11) already gives us the function ψ.

Sometimes it even suffices to know only the functions φ^Ω. If that is our goal, and if we do not yet have the finite transformations (18.9), we could directly start from

$$\frac{d\tilde{x}^n}{d\tilde{x}^1} = \frac{\xi^n}{\xi^1}, \qquad \frac{d\tilde{u}^\alpha}{d\tilde{x}^1} = \frac{\eta^\alpha}{\xi^1} \tag{18.13}$$

instead of (18.10) and thus again arrive at (18.12), the φ^Ω being the constants of integration arising when integrating (18.13).

We now have to hand all the information we need to reduce the number of variables by one and to go from the original system $H_A = 0$ of partial differential equations to the reduced system $G_A = 0$ described by equation (18.6). In detail, the procedure runs as follows. From the functions φ^Ω we choose an appropriate set of M functions $w^\alpha(x^i, u^\beta)$ as new dependent variables and take the rest as the $N-1$ new independent variables $y^k(x^i, u^\beta)$. Inverting the functions $\{\psi, y^k, w^\alpha\}$, we obtain $u^\alpha(\psi, y^k, w^\beta)$ and $x^n(\psi, y^k, w^\beta)$. We then assume that the w^β are functions only of the y^k, $w^\beta = w^\beta(y^k)$. To express the derivatives of the u^α appearing in $H_A = 0$, we use the chain rule, for example,

$$u^\alpha_{,n} = \frac{\partial u^\alpha}{\partial x^n} = \frac{\partial u^\alpha}{\partial w^\beta}\frac{\partial w^\beta}{\partial y^k}\frac{\partial y^k}{\partial x^n} + \frac{\partial u^\alpha}{\partial y^k}\frac{\partial y^k}{\partial x^n} + \frac{\partial u^\alpha}{\partial \psi}\frac{\partial \psi}{\partial x^n}, \tag{18.14}$$

where all partial derivatives except $\partial w^\beta/\partial y^k$ are known functions. If we insert everything into $H_A = 0$, the promise of the symmetry approach is that in the resulting system $G_A = 0$ all ψ's will disappear: only the y^k, the w^β, and the derivatives of the w^β will survive.

The following examples will show that this reduction procedure is less complicated and abstract than the reader may fear.

18.3 Examples

1. The wave equation

$$\Box u = u_{,xx} + u_{,yy} + u_{,zz} - u_{,tt} = 0 \tag{18.15}$$

admits (among others) the scaling invariance

$$\mathbf{X} = x\frac{\partial}{\partial x} + y\frac{\partial}{\partial y} + z\frac{\partial}{\partial z} + t\frac{\partial}{\partial t}. \tag{18.16}$$

This generator is so simple that ψ and the four similarity variables φ^Ω can easily be guessed from $\mathbf{X}\varphi^\Omega = 0$, $\mathbf{X}\psi = 1$. The result is, for example,

$$\varphi^1 = x/t, \qquad \varphi^2 = y/t, \qquad \varphi^3 = z/t, \qquad \varphi^4 = u, \qquad \psi = \ln t. \quad (18.17)$$

We take φ^4 as the new dependent variable $w\,(=u)$, and the remaining φ^k as new independent variables y^k, $k = 1, \ldots, 3$. To express the derivatives of u in the new variables, we use (18.14) and obtain, for example,

$$\begin{aligned} u_{,x} &= \frac{\partial u}{\partial w}\frac{\partial w}{\partial y^k}\frac{\partial y^k}{\partial x} = \frac{\partial w}{\partial y^k}\frac{\partial y^k}{\partial x} = \frac{\partial w}{\partial y^1}\frac{1}{t}, \\ u_{,t} &= \frac{\partial w}{\partial y^k}\frac{\partial y^k}{\partial t} = -\frac{\partial w}{\partial y^k}\frac{y^k}{t}. \end{aligned} \quad (18.18)$$

The t appearing here and also in the second derivatives of u will cancel out on inserting everything into the wave equation (18.15). The result is the differential equation

$$\frac{\partial^2 w}{\partial y^i \partial y^k}(\delta^{ik} - y^i y^k) - 2\frac{\partial w}{\partial y^i} y^i = 0, \quad (18.19)$$

which is independent of $\psi = \ln t$, and thus a differential equation in only three independent variables.

2. The heat conduction equation $u_{,xx} - u_{,t} = 0$ admits (among others) the symmetry with generator

$$\mathbf{X} = (a_1 + a_2 t)\frac{\partial}{\partial x} + \frac{\partial}{\partial t} - \frac{a_2}{2} x u \frac{\partial}{\partial u}. \quad (18.20)$$

Here we know already the finite transformations from (17.26): they are (with $a_3 = 1$)

$$\begin{aligned} \tilde{x} &= x + a_2 \varepsilon t + a_1 \varepsilon + \tfrac{1}{2} a_2 \varepsilon^2, \\ \tilde{u} &= u \exp(-\tfrac{1}{2} a_2 \varepsilon x - \tfrac{1}{4} a_2^2 \varepsilon^2 t - \tfrac{1}{4} a_1 a_2 \varepsilon^2 - \tfrac{1}{12} a_2^2 \varepsilon^3), \end{aligned} \quad (18.21)$$

and

$$\tilde{t} = t + \varepsilon. \quad (18.22)$$

So we can use these finite transformations to construct ψ and the similarity variables φ^Ω. According to the general scheme, this will be done by inverting one of the equations to get $\varepsilon = \psi$, and then inserting this into the rest of the equations. We choose (18.22) to obtain $\varepsilon = \tilde{t} - t$. Inserting this expression for ε into (18.21) yields

$$\tilde{x} = a_1\tilde{t} + \tfrac{1}{2}a_2\tilde{t}^2 + x - a_1 - \tfrac{1}{2}a_2t^2,$$
$$\tilde{u} = u\exp[-\tfrac{1}{3}a_2^2\tilde{t}^3 - \tfrac{1}{4}a_1a_2\tilde{t}^2 - \tfrac{1}{2}a_2\tilde{t}(x - a_1 - \tfrac{1}{2}a_2t^2)$$
$$-\tfrac{1}{6}a_2^2t^3 - \tfrac{1}{4}a_1a_2t^2 + \tfrac{1}{2}a_2xt]. \tag{18.23}$$

Comparing this with the general form (18.12) and keeping in mind that $\tilde{x}^1$ there is $\tilde{t}$ here, we can read off φ^Ω as

$$\varphi^1 = x - a_1 - \tfrac{1}{2}a_2t^2,$$
$$\varphi^2 = u\exp(\tfrac{1}{2}a_2xt - \tfrac{1}{4}a_1a_2t^2 - \tfrac{1}{6}a_2^2t^3). \tag{18.24}$$

Of course, we choose φ^2 as the new dependent variable w and φ^1 as the independent variable y. If we now ask for solutions $w = w(y)$, we have,

$$u = w(y)\exp(-\tfrac{1}{2}a_2xt + \tfrac{1}{4}a_1a_2t^2 + \tfrac{1}{6}a_2^2t^3),$$
$$u_{,x} = [w_{,y} - \tfrac{1}{2}a_2tw]\exp(-\tfrac{1}{2}a_2xt + \tfrac{1}{4}a_1a_2t^2 + \tfrac{1}{6}a_2^2t^3), \tag{18.25}$$

and so on, and the heat conduction equation in the variables $w = \varphi^2$ and $y = \varphi^1$ reads

$$w_{,yy} + a_1w_{,y} + \tfrac{1}{2}a_2yw = 0, \tag{18.26}$$

so it is an ordinary differential equation as it should be – x and t cancel out.

3. Any invariant of a group of symmetries is a similarity variable of any of its generators. So since we know that in the case of the differential equation (16.70) the function (17.30), that is,

$$w = Px^{-3/2}, \tag{18.27}$$

is invariant under the action of any of the symmetry generators

$$\mathbf{X} = (ax + bxy)\frac{\partial}{\partial x} + [ay + \tfrac{1}{2}b(y^2 - x^2) + c]\frac{\partial}{\partial y} + \tfrac{3}{2}(aP + byP)\frac{\partial}{\partial P}, \tag{18.28}$$

we know that w is one of the two similarity variables. To get the second, we start from (18.13), that is, from

$$\frac{\mathrm{d}y}{\mathrm{d}x} = \frac{\xi^2}{\xi^1} = \frac{ay + b(y^2 - x^2)/2 + c}{ax + bxy}. \tag{18.29}$$

This equation can be written as

$$0 = \frac{a+by}{x}\,\mathrm{d}y - \frac{ay + b(y^2 - x^2)/2 + c}{x^2}\,\mathrm{d}x$$

$$= \mathrm{d}\left(\frac{c + b(x^2 + y^2)/2 + ay}{x}\right), \tag{18.30}$$

which shows that

$$z = \frac{1}{x}[c + \tfrac{1}{2}b(x^2 + y^2) + ay] \tag{18.31}$$

is constant along the integral curves of (18.29) and can be taken as the second similarity variable. We do not need the function ψ (with $\mathbf{X}\psi = 1$); w and z are sufficient to rewrite the differential equation (16.70). The final result is

$$ww_{,zz}(z^2 + a^2 - 2bc) + 2zww_{,z} - (z^2 + a^2 - 2bc){w^2}_{,z} - \tfrac{3}{2}w^2$$
$$+ 3\sqrt{2} = 0, \tag{18.32}$$

an ordinary second order differential equation. Note that although the symmetry generator (18.28) depends on three parameters a, b, and c, the differential equation (18.32) contains only one parameter $a^2 - 2bc$.

4. The two-dimensional potential equation $u_{,\zeta\bar{\zeta}} = 0$ admits the symmetry generator

$$\mathbf{X} = B(\zeta)\frac{\partial}{\partial\zeta} + \bar{B}(\bar{\zeta})\frac{\partial}{\partial\bar{\zeta}} + [W(\zeta) + \bar{W}(\bar{\zeta})]\frac{\partial}{\partial u}, \tag{18.33}$$

which depends on the two arbitrary functions $B(\zeta)$ and $W(\zeta)$ and their complex conjugates. If we ask for similarity variables, we have to think of these functions as given, fixed functions. To find the two similarity variables in the case $W \neq 0$, we can proceed as follows.

First we make a conformal transformation

$$\hat{\zeta} = f(\zeta), \qquad f' = \frac{1}{B(\zeta)\sqrt{2}}, \qquad \hat{\xi} = \frac{1}{\sqrt{2}}(v + ir) \tag{18.34}$$

to arrive at the simple form

$$\mathbf{X} = \frac{\partial}{\partial v} + h(v, r)\frac{\partial}{\partial u}, \qquad h_{,vv} + h_{,rr} = 0 \tag{18.35}$$

of the generator. Now the two similarity variables can be determined as the integration parameters (constants) of the system (18.13), that is, of

$$\frac{\mathrm{d}r}{\mathrm{d}v}=0, \qquad \frac{\mathrm{d}u}{\mathrm{d}v}=h(v,r). \tag{18.36}$$

They turn out to be

$$\varphi=r, \qquad w=u-\int_{v_0}^{v} h(v,\varphi)\,\mathrm{d}v, \tag{18.37}$$

and in these variables the potential equation $u_{,vv}+u_{,rr}=0$ reads (if $h_{,vv}+h_{,rr}=0$ is taken into account)

$$w_{,\varphi\varphi}=-h_{,v}(v_0,\varphi). \tag{18.38}$$

So if we start from any given solution $h(v,r)$ of the potential equation, then a function $w(\varphi)$ can be determined from (18.38) such that

$$u(v,r)=\int_{v_0}^{v} h(v,r)\,\mathrm{d}v+w(r) \tag{18.39}$$

is again a solution.

5. The symmetry generator of Einstein's field equations (16.68) is

$$\mathbf{X}=\xi^k(x^n)\frac{\partial}{\partial x^k}-(\xi^k{}_{,i}g_{kn}+\xi^k{}_{,n}g_{ik}+ag_{in})\frac{\partial}{\partial g_{in}}, \tag{18.40}$$

where the $\xi^k(x^n)$ are arbitrary functions. We shall not try to carry out the necessary calculations to find the reduced form of the field equations but only set out to obtain some information about the similarity solutions. For this we shall not use the general approach but instead investigate the equation (18.7), that is,

$$\eta^\alpha-\xi^k u^\alpha{}_{,k}=0, \tag{18.41}$$

which has to be satisfied by every similarity solution.

For the generator (18.40), this equation reads

$$\xi^k g_{in,k}+\xi^k{}_{,i}g_{kn}+\xi^k{}_{,n}g_{ik}+ag_{in}=0, \tag{18.42}$$

which can be written as

$$\xi_{i;n} + \xi_{n;i} = \mathscr{L}_{\xi} g_{in} = -a g_{in}; \tag{18.43}$$

compare Sections 14.4 and 16.1.

So a metric g_{in} is a similarity solution with $a = 0$ exactly if space-time admits a Killing vector with components ξ^n and with $a \neq 0$ exactly if the ξ^n are the components of a homothetic vector (in both cases, of course, Einstein's field equations also have to be satisfied). This fact explains the outstanding role that Killing vectors and homothetic vectors play in the search for exact solutions.

By a coordinate transformation $\hat{x}^n = \hat{x}^n(x^i)$, the vector ξ^k can always be transformed to $\hat{\xi}^k = (1, 0, 0, 0)$. In these coordinates, (18.42) yields

$$\frac{\partial \hat{g}_{in}}{\partial \hat{x}^1} = -a \hat{g}_{in}, \tag{18.44}$$

from which

$$\hat{g}_{in} = e^{-a\hat{x}_1} h_{in}(\hat{x}^2, \hat{x}^3, \hat{x}^4) \tag{18.45}$$

follows. This is just another way of saying that $\mathbf{X}$ can be transformed to $\mathbf{X} = \partial/\partial\hat{x}^1$ and that $\{\hat{x}^2, \hat{x}^3, \hat{x}^4, h_{in}\}$ are the similarity variables.

Obviously the similarity solutions of Einstein's equations are space-times with the property that the geodesic equation has linear first integrals; compare Section 13.4.

18.4 Conditional symmetries

The existence of a symmetry ensures the possibility of introducing similarity variables and thus of a reduction of the number of variables. But the converse is not true. Take, for instance, the differential equation

$$\frac{1}{r^2}\frac{\partial}{\partial r}\left(r^2 \frac{\partial u}{\partial r}\right) + \frac{1}{r^2 \sin\vartheta}\frac{\partial}{\partial\vartheta}\left(\sin\vartheta \frac{\partial u}{\partial\vartheta}\right) - \frac{\partial^2 u}{\partial t^2} = 0, \tag{18.46}$$

which is a special case of the wave equation in spherical coordinates. Clearly $\mathbf{X} = \partial/\partial\vartheta$ is not a symmetry of this equation, nor is any generator of the form $f(\vartheta)\,\partial/\partial\vartheta$, but nevertheless solutions $u = u(r, t)$ exist, solutions that are independent of ϑ.

One can subsume those cases under symmetries if one introduces the notion of a *conditional symmetry*. If – possibly only after an appropriate transformation of variables $\{x^i, u^\alpha\} \to \{\psi, y^k, w^\alpha\}$ – solutions $w^\alpha(y^k)$ of a system $H_A(x^i, u^\alpha, u^\alpha{}_{,n}, \ldots) = 0$ of differential equations exist that are independent of the variable ψ, then $\mathbf{X} = \partial/\partial\psi$ is a symmetry under the condition that it is applied only to solutions independent of ψ and is thus called a conditional symmetry of $H_A = 0$.

This rather trivial definition and notion makes sense if it can be expressed in invariant form and if it can then be used to find those special classes of solutions or at least the differential equations governing them. That is indeed possible.

The invariant form can easily be obtained. We know already that a solution surface $u^\alpha - u^\alpha(x^i) = 0$ is invariant under a generator $\mathbf{X}$ if

$$\eta^\alpha - \xi^n u^\alpha{}_{,n} = 0 \tag{18.47}$$

holds. Since this equation is an identity in the variables x^i, its derivatives with respect to the x^i will also vanish. So we obtain

$$\frac{\mathrm{D}}{\mathrm{D}x^i}(\eta^\alpha - \xi^n u^\alpha{}_{,n}) = \frac{\mathrm{D}\eta^\alpha}{\mathrm{D}x^i} - u^\alpha{}_{,n}\frac{\mathrm{D}\xi^n}{\mathrm{D}x^i} - u^\alpha{}_{,ni}\xi^n = 0, \tag{18.48}$$

or – because of the definition (15.15) of the η^α_i –

$$\eta^\alpha_i - \xi^n u^\alpha{}_{,ni} = 0, \tag{18.49}$$

and similarly for the higher derivatives. In variables with $\mathbf{X} = \partial/\partial\psi$ and $u^\alpha = w^\alpha(y^k)$, equation (18.47) amounts to $\partial w/\partial\psi = 0$ and (18.49) to $\partial^2 w/\partial\psi\,\partial y^i = 0$. A conditional symmetry of a system $H_A = 0$ is then a symmetry with generator

$$\mathbf{X} = \xi^i \frac{\partial}{\partial x^i} + \eta^\alpha \frac{\partial}{\partial u^\alpha} + \eta^\alpha_i \frac{\partial}{\partial u^\alpha{}_{,i}} + \cdots, \tag{18.50}$$

which has the property

$$\mathbf{X}H_A(x^i, u^\alpha, u^\alpha{}_{,i}, \ldots) = 0 \qquad (\text{mod}\, H_A = 0) \tag{18.51}$$

only for those solutions for which (18.47) *and its consequences* [*e.g.*, (18.49)] *hold.*

Symmetries are included here as those conditional symmetries for which (18.51) is true without making use of (18.47) and its consequences. From the computational point of view, the main difference between symmetries and conditional symmetries is that, when $H_A = 0$ has been properly used, the equations $\mathbf{X}H_A = 0$ are identities in all remaining variables in the case of symmetries, whereas for conditional symmetries some more of the $u^{\alpha}{}_{,i}$ and their derivatives have to be substituted by means of equations (18.47), (18.49), and so on, and only then can $\mathbf{X}H_A = 0$ be treated as identities in the surviving set of variables.

It is clear that there are more conditional symmetries than symmetries. But as in most cases when we weaken the conditions to be satisfied by a particular class of symmetries, the price we have to pay is that we are no longer able to determine the now larger class of symmetries in practice. Conditional symmetries do not play an important role mainly due to these difficulties.

To give at least one example of a conditional symmetry and its determination, we take the heat conduction equation

$$u_{,xx} - u_{,t} = 0 \tag{18.52}$$

and ask for conditional symmetries of the restricted form

$$\mathbf{X} = \sigma(x, t, u)\frac{\partial}{\partial x} + \frac{\partial}{\partial t}. \tag{18.53}$$

Equation (18.47) here reads

$$\sigma u_{,x} + u_{,t} = 0, \tag{18.54}$$

and its first derivatives give, if (18.52) and (18.54) are taken into account,

$$\begin{aligned} u_{,xt} &= \left(\frac{\sigma_{,x}}{\sigma} - \sigma\right)u_{,t} - \frac{\sigma_{,u}}{\sigma^2}u_{,t}^2, \\ u_{,tt} &= \left(\frac{\sigma_{,t}}{\sigma} + \sigma^2 - \sigma_{,x}\right)u_{,t} + 2\frac{\sigma_{,u}}{\sigma}u_{,t}^2. \end{aligned} \tag{18.55}$$

Equations (18.52), (18.54), and (18.55) show that all first and second derivatives of u can be expressed in terms of $u_{,t}$ (and of σ and its derivatives).

If we now write down the symmetry condition $\mathbf{X}H = \eta_{xx} - \eta_t = 0$ in full by inserting the explicit expressions for η_{xx} and η_t, we obtain

$$\begin{aligned} &-\sigma_{,xx}u_{,x} - 2\sigma_{,ux}u_{,x}^2 - \sigma_{,uu}u_{,x}^3 - 2\sigma_{,x}u_{,xx} - 3\sigma_{,u}u_{,x}u_{,xx} \\ &\quad + \sigma_{,t}u_{,x} + \sigma_{,u}u_{,x}u_{,t} = 0. \end{aligned} \tag{18.56}$$

Substituting $u_{,xx}$ and $u_{,x}$ by means of (18.52) and (18.54) yields

$$\frac{\sigma_{,uu}}{\sigma^3}u_{,t}^3 + 2\left(\frac{\sigma_{,u}}{\sigma} - \frac{\sigma_{,ux}}{\sigma}\right)u_{,t}^2 + \left(\frac{\sigma_{,xx}}{\sigma} - 2\sigma_{,x} - \frac{\sigma_{,t}}{\sigma}\right)u_{,t} = 0. \tag{18.57}$$

Since now as many derivatives of u as possible have been eliminated, this equation is an identity in $u_{,t}$. Equating to zero the coefficients of the different powers of $u_{,t}$, we see that the function σ is independent of u and must satisfy

$$\sigma_{,xx} - 2\sigma\sigma_{,x} - \sigma_{,t} = 0, \qquad \sigma = \sigma(x, t). \tag{18.58}$$

Disappointingly enough, to find conditional symmetries, we must solve a nonlinear partial differential equation more complicated than the differential equation we hoped to solve with those symmetries!

Special solutions to (18.58) can easily be found. The solution

$$\sigma = \frac{a_1 + a_2 x}{a_5 + 2a_2 t}, \qquad a_i = \text{const.}, \tag{18.59}$$

corresponds to one of the ordinary symmetries (16.64); compare Exercise 18.5. Another simple solution, not equivalent to an ordinary symmetry, is

$$\sigma = \tan x, \qquad \mathbf{X} = \tan x \frac{\partial}{\partial x} + \frac{\partial}{\partial t}. \tag{18.60}$$

The "similarity" variables belonging to this generator are u and

$$\varphi = e^{-t} \sin x. \tag{18.61}$$

If we write the heat conduction equation in these variables, we obtain

$$u_{,\varphi\varphi} = 0, \qquad u = a\varphi + b,$$

where φ is a special solution of the heat conduction equation, and choosing it as a ("similarity") variable does not lead to anything new.

18.5 Exercises

1 Find the similarity variables of the generator $\mathbf{X} = a\,\partial/\partial v + x\,\partial/\partial x + y\,\partial/\partial y$ and use them to write the system (16.30) as a system of ordinary differential equations in the independent variable x/y.

2 Is $\mathbf{X} = \partial/\partial x$ a symmetry and x a similarity variable of the differential equation $x^2(u_{,xx} + u_{,y}) = 0$?

3 Find the similarity variables of the Galilei transformation $\mathbf{X} = 6t\,\partial/\partial x + \partial/\partial u$ and use them to reduce the Korteweg–deVries equation (16.72).

4 Show that $\eta^\alpha - \xi^n u^\alpha_{,n} = 0$ is invariant under the operator $\mathbf{X}$ to which it belongs.

5 Show that every solution $h(x, t)$ of the heat conduction equation can be chosen as a similarity variable of a suitable conditional symmetry (18.53). How are h and σ related to each other?

6 Show that if $\mathbf{X}$ is the generator of a conditional Lie point symmetry, then the same is true for $\hat{\mathbf{X}} = \lambda(x^i, u^\alpha)\mathbf{X}$ for arbitrary functions λ. (*Hint*: the extension law is what matters!)

7 Check that $\mathbf{Y} = \partial/\partial\vartheta$ is a conditional symmetry of (18.46).

19

How to use Lie point symmetries of partial differential equations III: Multiple reduction of variables and differential invariants

19.1 Multiple reduction of variables step by step

In the preceding chapter we learned how to reduce the number of independent variables by one if the partial differential equations in question admit a Lie point symmetry and how to establish the differential equations for this special set of solutions.

Differential equations can admit more than one symmetry, so that we have several possible ways to choose a set of similarity variables by starting from different symmetries. Can we also use the existence of a multiple-parameter group of symmetries to achieve a multiple reduction of variables? The answer is yes, and there are two main roads to doing this. The first way, going step by step, will be discussed immediately. The second way, which uses invariants, will be presented in the subsequent section.

Reducing the number of variables step by step obviously means first taking a symmetry $\mathbf{X}$, introducing its similarity variables $\varphi^{\Omega} = \{y^k, w^\alpha\}$ defined by

$$\mathbf{X}\varphi^{\Omega} = 0, \qquad \mathbf{X} \equiv \partial/\partial\psi, \tag{19.1}$$

and writing the original system $H_A = 0$ of differential equations as

$$G_A(y^k, w^\alpha, w^\alpha{}_{,k}, \ldots) = 0, \tag{19.2}$$

and then using a second symmetry of $H_A = 0$ to accomplish a further reduction, and so on.

This idea will work only if the symmetry $\mathbf{Y}$ is really inherited by $G_A = 0$; that is, if from $\mathbf{Y}H_A = 0 \pmod{H_A = 0}$

$$\mathbf{Y}G_A = 0 \qquad (\mathrm{mod}\, G_A = 0) \tag{19.3}$$

follows. This is not necessarily the case since although the finite transformations generated by $\mathbf{Y}$ will map solutions of $H_A = 0$ into solutions of $H_A = 0$, they will not in general map solutions of $G_A = 0$ into solutions of $G_A = 0$: the transformed solutions can depend on ψ even if the original solutions did not. To generate solutions independent of ψ (which then satisfy $G_A = 0$), the finite transformations of the w^α and y^n induced by $\mathbf{Y}$ must be independent of ψ so that for this generator

$$\mathbf{Y} = \mu(\psi, y^n, w^\beta)\frac{\partial}{\partial\psi} + \xi^k(y^n, w^\beta)\frac{\partial}{\partial y^k} + \eta^\alpha(y^n, w^\beta)\frac{\partial}{\partial w^\alpha} \tag{19.4}$$

must hold (the dependence of μ on ψ does not matter, as it enters only into the finite transformations of ψ, which do not affect the ψ-independent solutions of $G_A = 0$). Because $\mathbf{X} = \partial/\partial\psi$, an invariant criterion for $\mathbf{Y}$ to have this property is

$$[\mathbf{X}, \mathbf{Y}] = \lambda(\psi, y^k, w^\alpha)\mathbf{X}, \tag{19.5}$$

where λ is an arbitrary function of all variables. If $\mathbf{Y}$ satisfies such a commutator relation, then the reduced set $G_A = 0$ inherits the symmetry $\mathbf{Y}$.

This result is in close analogy to that obtained in Section 9.2 when asking for the possibilities that the existence of several symmetries offers for the integration of ordinary differential equations. The reduction of the *number* of independent variables here is analogous to the reduction of the *order* of the differential equation there. It may even happen that a symmetry of a partial differential equation is inherited by the ordinary differential equation arising from it when introducing similarity variables.

Note that the reduced system $G_A = 0$ may also have symmetries that do *not* originate in symmetries of the original $H_A = 0$!

We now want to illustrate all this by a few examples. As a *first example* we consider the wave equation

$$\Box u = u_{,xx} + u_{,yy} + u_{,zz} - u_{,tt} = 0. \tag{19.6}$$

As shown in Section 18.3, the scaling symmetry $\mathbf{X} = x\,\partial/\partial x + y\,\partial/\partial y + z\,\partial/\partial z + t\,\partial/\partial t$ can be used to transform (reduce) it to

$$\frac{\partial^2 w}{\partial y^i\,\partial y^k}(\delta^{ik} - y^i y^k) - 2y^i\frac{\partial w}{\partial y^i} = 0, \tag{19.7}$$

$$y^1 = x/t, \qquad y^2 = y/t, \qquad y^3 = z/t, \qquad w = u.$$

The generator

$$\mathbf{Y} = y\frac{\partial}{\partial x} - x\frac{\partial}{\partial y} \tag{19.8}$$

of a rotation in the x–y plane is a second symmetry of the wave equation. It commutes with **X**, and so

$$\mathbf{Y} = Y(y^1)\frac{\partial}{\partial y^1} + Y(y^2)\frac{\partial}{\partial y^2} = y^2\frac{\partial}{\partial y^1} - y^1\frac{\partial}{\partial y^2} \tag{19.9}$$

is a symmetry of (19.7). Its obvious similarity variables are w, y^3, and $v = (y^1)^2 + (y^2)^2 = (x^2 + y^2)/t^2$,

$$\mathbf{Y}w = \mathbf{Y}v = \mathbf{Y}s = 0, \qquad v \equiv (y^1)^2 + (y^2)^2, \qquad s \equiv y^3, \tag{19.10}$$

so that solutions $w = w(s, v)$ may exist that then obey

$$\begin{aligned} &4v(1-v)w_{,vv} - 4vsw_{,vs} - (1-s^2)w_{,ss} \\ &\quad + (4-6v)w_{,v} - 2sw_{,s} = 0. \end{aligned} \tag{19.11}$$

Among the many symmetries of the wave equation, there is still another that commutes with both **X** and **Y** and thus offers the possibility of a further reduction. This is the generator

$$\mathbf{Z} = t\frac{\partial}{\partial z} + z\frac{\partial}{\partial t} \tag{19.12}$$

of a Lorentz transformation (pseudorotation). In variables w, s, and v it reads

$$\mathbf{Z} = (\mathbf{Z}s)\frac{\partial}{\partial s} + (\mathbf{Z}v)\frac{\partial}{\partial v} + (\mathbf{Z}w)\frac{\partial}{\partial w} = (1-s^2)\frac{\partial}{\partial s} - 2vs\frac{\partial}{\partial v}. \tag{19.13}$$

Its similarity variables are w and

$$\sigma = v/(1-s^2) = (x^2+y^2)/(t^2-z^2), \tag{19.14}$$

and in these variables the wave equation is given as

$$\sigma w_{,\sigma\sigma} + w_{,\sigma} = 0. \tag{19.15}$$

This ordinary differential equation still has the symmetry with generator $\mathbf{V} = \partial/\partial w$ but also a symmetry not inherited from the wave equation, namely, $\mathbf{W} = \sigma\,\partial/\partial\sigma$. So it can easily be integrated, yielding the particular one- (independent-) variable solution

$$u = w = a_1 + a_2 \ln \sigma = a_1 + a_2 \ln \frac{x^2 + y^2}{t^2 - z^2} \tag{19.16}$$

of the four-variable wave equation.

As a *second example* we take the differential equation (16.70), that is,

$$P(P_{,xx} + P_{,yy}) - (P^2{}_{,x} + P^2{}_{,y}) + 3x\sqrt{2} = 0, \tag{19.17}$$

which admits the three symmetries

$$\mathbf{X}_1 = x\frac{\partial}{\partial x} + y\frac{\partial}{\partial y} + \tfrac{3}{2}P\frac{\partial}{\partial P},$$

$$\mathbf{X}_2 = xy\frac{\partial}{\partial x} + \tfrac{1}{2}(y^2 - x^2)\frac{\partial}{\partial y} + \tfrac{3}{2}yP\frac{\partial}{\partial P},$$

$$\mathbf{X}_3 = \frac{\partial}{\partial y}. \tag{19.18}$$

We know from Example 3 of Section 18.3 that the similarity variables belonging to $\mathbf{X} = a\mathbf{X}_1 + b\mathbf{X}_2 + c\mathbf{X}_3$ are $w = Px^{-3/2}$ and $z = [c + b(x^2 + y^2)/2 + ay]/x$ and that (19.17) can be reduced to the ordinary differential equation (18.32).

It is easy to show that a generator

$$\mathbf{Y} = A\mathbf{X}_1 + B\mathbf{X}_2 + C\mathbf{X}_3 \tag{19.19}$$

that obeys the condition $[\mathbf{X}, \mathbf{Y}] = \lambda\mathbf{X}$ will exist only if $a^2 - 2bc = 0$. That means that from the one-parameter set (18.32) of ordinary differential equations only

$$z^2 ww_{,zz} + 2zww_{,z} - z^2 w^2{}_{,z} - 3w^2/2 + 3\sqrt{2} = 0 \tag{19.20}$$

inherits a symmetry $\mathbf{Y}$ that then turns out to be

$$\mathbf{Y} = z\frac{\partial}{\partial z}. \tag{19.21}$$

So (19.20) can be reduced to a first order differential equation [but cannot be readily solved by quadratures since no further symmetry is available: the three symmetries (19.18) span only a two-dimensional space].

As a *third example* we consider the wave equation in spherical coordinates,

$$\square u = \frac{1}{r^2}\frac{\partial}{\partial r}\left(r^2\frac{\partial u}{\partial r}\right) + \frac{1}{r^2 \sin\vartheta}\frac{\partial}{\partial\vartheta}\left(\sin\vartheta\frac{\partial u}{\partial\vartheta}\right) + \frac{1}{r^2\sin^2\vartheta}\frac{\partial^2 u}{\partial\varphi^2} - \frac{\partial^2 u}{\partial t^2} = 0. \tag{19.22}$$

Among other symmetries, it admits the group of three-dimensional rotations generated by

$$\mathbf{X}_1 = \frac{\partial}{\partial\varphi}, \qquad \mathbf{X}_2 = \sin\varphi\frac{\partial}{\partial\vartheta} + \cos\varphi\cot\vartheta\frac{\partial}{\partial\varphi},$$
$$\mathbf{X}_3 = \cos\varphi\frac{\partial}{\partial\vartheta} - \sin\varphi\cot\vartheta\frac{\partial}{\partial\varphi}. \tag{19.23}$$

If we start the reduction procedure with $\mathbf{X} = \mathbf{X}_1 = \partial/\partial\varphi$, then $\{u, r, \vartheta, t\}$ are the similarity variables and the wave equation reduces to

$$\frac{1}{r^2}\frac{\partial}{\partial r}\left(r^2\frac{\partial u}{\partial r}\right) + \frac{1}{r^2\sin\vartheta}\frac{\partial}{\partial\vartheta}\left(\sin\vartheta\frac{\partial u}{\partial\vartheta}\right) - \frac{\partial^2 u}{\partial t^2} = 0. \tag{19.24}$$

Surprisingly, this equation does not inherit either of the two symmetries $\mathbf{X}_2$ and $\mathbf{X}_3$ (or any combination of them) because $[\mathbf{X}, \mathbf{Y}] = \lambda\mathbf{X}$ cannot be satisfied for any $\mathbf{Y} = a_2\mathbf{X}_2 + a_3\mathbf{X}_3$. Hence $\mathbf{X}_2$ and $\mathbf{X}_3$ cannot be used for a further reduction by simply going step by step.

The surprise in this result is due to the fact that, of course, the rotational symmetry of the wave equation can be used to construct solutions that are independent of both the coordinates φ *and* ϑ; solutions $u = u(r, t)$ that then obey

$$\frac{1}{r^2}\frac{\partial}{\partial r}\left(r^2\frac{\partial u}{\partial r}\right) - \frac{\partial^2 u}{\partial t^2} = 0 \tag{19.25}$$

exist! At first glance one might suspect a contradiction here, but there is none: as shown in Section 18.4, $\mathbf{Y} = \partial/\partial\vartheta$ is only a conditional symmetry of (19.24), and conditional symmetries are not covered by $[\mathbf{X}, \mathbf{Y}] = \lambda\mathbf{X}$.

Have we really to include conditional symmetries, that is, symmetries that are difficult to detect and handle, to be able to cover all possible multiple reductions? The answer is no, there are better ways to exploit the existence of a multiple-parameter group of symmetries, and we shall now explain how this can be done.

19.2 Multiple reduction of variables by using invariants

As already discussed in Section 6.4, invariants of a given group with generators $\mathbf{X}_A$,

$$\mathbf{X}_A = \underset{A}{\xi^i}(x^n, u^\beta)\frac{\partial}{\partial x^i} + \underset{A}{\eta^\alpha}(x^n, u^\beta)\frac{\partial}{\partial u^\alpha}, \tag{19.26}$$

are solutions $\varphi^\Omega(x^n, u^\beta)$ to the equations

$$\mathbf{X}_A\varphi^\Omega = 0, \qquad A = 1,\ldots,r. \tag{19.27}$$

The orbits of the group can be given in terms of these invariants as $\varphi^\Omega = \text{const.}$

If we think of the $\mathbf{X}_A$ as the extensions up to the pth derivative of (19.26),

$$\mathbf{X}_A = \mathbf{X}_A^{(p)} = \underset{A}{\xi^i}\frac{\partial}{\partial x^i} + \underset{A}{\eta^\alpha}\frac{\partial}{\partial u^\alpha} + \underset{A}{\eta^\alpha_i}\frac{\partial}{\partial u^\alpha{}_{,i}} + \cdots + \underset{A}{\eta^\alpha_{n_1\cdots n_p}}\frac{\partial^p}{\partial u^\alpha{}_{,n_1\cdots n_p}}, \tag{19.28}$$

that is, as acting in the space of variables $\{x^i, u^\alpha, u^\alpha{}_{,n}, \ldots, u^\alpha{}_{,n_1\ldots n_p}\}$, then the solutions to (19.27) are called differential invariants. For a given group, invariants need not always exist since the number k of linearly independent generators, that is, the number of linearly independent differential equations (19.27), can be larger than (or equal to) the dimension of the (x^n, u^α)-space. But by going to higher and higher derivatives, the dimension of the representation space can be made arbitrarily large, larger than the dimension of the space spanned by the r vectors $\mathbf{X}_A^{(p)}$: differential invariants of sufficiently high order always exist.

Differential invariants frequently appear as building blocks of differential equations, but in most cases they are of no direct use for an integration strategy for partial differential equations: it is the invariants that are important here.

In fact we already used invariants when performing reduction of variables: the similarity variables introduced in Section 18.2 and further exploited in the foregoing section are nothing other than invariants! So if we already used invariants, what is the better way of doing this we promised to show?

The answer is that we should not determine the invariants (similarity variables) of a first symmetry $\mathbf{X}$, then try to find a second symmetry $\mathbf{Y}$

satisfying $[\mathbf{X}, \mathbf{Y}] = \lambda \mathbf{X}$, and again determine its invariants, but instead we should find the invariants of the group generated by **X** and **Y** in *one* go!

More generally, if we have a group of symmetries with generators $\mathbf{X}_A$, and if we want to achieve a maximal reduction of variables, we should determine its invariants φ^Ω, divide them into dependent variables w^α and independent variables y^k, and derive the differential equations $G_A = 0$ for $w^\alpha(y^k)$ from the original equations $H_A = 0$ by using the chain rule. If there are no invariants of the complete group, or if we are interested in a less than maximal reduction, we have to use a subgroup – at least a one-dimensional subgroup will always exist!

To find the invariants φ^Ω, we have either to solve the system (19.27) of r differential equations or – if we already know the finite transformations of the group – to eliminate the r group parameters from these finite transformations; compare Section 18.2, where this was explained for a one-parameter group. If necessary, we can introduce coordinates (that generalize the ψ of Section 18.2) to label the points on each orbit $\varphi^\Omega = \text{const}$. But in most cases we shall not need them.

We do not want to treat all technical details here but instead illustrate the above idea by some examples.

If we want to use the full rotation group (19.23) for a multiple reduction of the wave equation (19.22), we need the invariant of that group in the five-dimensional $(r, \vartheta, \varphi, t, u)$-space. Since the orbits are two-dimensional (the three $\mathbf{X}_A$ are linearly dependent because $\cot \vartheta \mathbf{X}_1 + \sin \varphi \mathbf{X}_3 - \cos \varphi \mathbf{X}_2 = 0$), there are exactly three invariants, namely, u, r, and t, and they lead directly to (19.25).

If we intend to use the two symmetries (19.8) and (19.12), that is,

$$\mathbf{X}_1 = y\frac{\partial}{\partial x} - x\frac{\partial}{\partial y}, \qquad \mathbf{X}_2 = t\frac{\partial}{\partial z} + z\frac{\partial}{\partial t}, \tag{19.29}$$

for the reduction of the wave equation in Cartesian coordinates, then we need the invariants satisfying

$$\mathbf{X}_1 \varphi^\Omega = 0 = \mathbf{X}_2 \varphi^\Omega. \tag{19.30}$$

They are easily recognized to be

$$\varphi^1 = u, \qquad \varphi^2 = x^2 + y^2, \qquad \varphi^3 = z^2 - t^2, \tag{19.31}$$

and because of

$$u_{,x} = \frac{\partial u}{\partial \varphi^2}\frac{\partial \varphi^2}{\partial x} + \frac{\partial u}{\partial \varphi^3}\frac{\partial \varphi^3}{\partial x} = 2x\frac{\partial u}{\partial \varphi^2} \equiv 2xu_{,2}, \quad \text{etc.}, \tag{19.32}$$

the reduced wave equation becomes

$$\varphi^2 u_{,22} + \varphi^3 u_{,33} + u_{,2} + u_{,3} = 0. \tag{19.33}$$

An interesting example is provided by the differential equation (19.17) and its three symmetries (19.18). In the three-dimensional (x, y, P)-space, the orbits of the group are two-dimensional – there is only one invariant, namely, $w = Px^{-3/2}$. The same is true for any two-parameter subgroup of (19.18). But for writing down a differential equation, we need at least two invariants – so, apparently, neither the full symmetry nor any of its two-dimensional subgroups can be used for a multiple reduction. So why did we succeed in the preceding section in performing a twofold reduction step by step, first to an ordinary second order differential equation (19.29) and then – in principle – to a first order equation by using the symmetry (19.21)? On which invariants is this reduction based? The answer to this question is that when actually performing the second reduction, it is a reduction in order, and not in number of variables, and the new variable we have to introduce is essentially the derivative of w (as a function of w), that is, we use a *differential* invariant!

19.3 Some remarks on group-invariant solutions and their classification

The solutions we shall find when performing a reduction of variables by means of a particular group are of course invariant under this group simply because all variables satisfy $\mathbf{X}_A\varphi^\Omega = 0$. For the particular case of a one-parameter group, we called those solutions *similarity solutions*. To stress their invariance property, they are also called *group-invariant solutions*.

To draw the maximum profit out of an existing group of symmetries, one certainly should determine as many group-invariant solutions as possible; that is, one should use as many subgroups as possible for a reduction of variables. This last statement sounds plausible, but it is not quite true: one should rather say that the goal is to use as many subgroups as necessary.

To understand the problem, let us consider the three-dimensional potential equation

$$u_{,xx} + u_{,yy} + u_{,zz} = 0 \tag{19.34}$$

and the group with generators

$$\mathbf{X}_1 = \frac{\partial}{\partial x}, \quad \mathbf{X}_2 = \frac{\partial}{\partial y}, \quad \mathbf{X}_3 = x\frac{\partial}{\partial y} - y\frac{\partial}{\partial x},$$

$$\mathbf{X}_4 = \frac{\partial}{\partial z}, \quad \mathbf{X}_5 = \frac{\partial}{\partial u}, \quad \mathbf{X}_6 = u\frac{\partial}{\partial u}, \tag{19.35}$$

corresponding to translations and rotations in the x–y plane, translations in the z- and u-directions, and the multiplication of u by a constant factor. If we ask for one-dimensional z-independent solutions, that is, solutions invariant under $\mathbf{X}_4$ and any one of the $\{\mathbf{X}_1, \mathbf{X}_2, \mathbf{X}_3\}$ or a linear combination of the latter, we shall find, for example,

$$\begin{aligned} u &= ax + c, \\ u &= by + d, \\ u &= ex + fy + g. \end{aligned} \qquad (19.36)$$

Of course, these solutions are, strictly speaking, all different. But one certainly feels that they differ only trivially: they can be transformed into each other by adding constants, multiplying with constants, and performing rotations in the x–y plane. That is, all these solutions (19.36) can be obtained from, for example, $u = x$ by performing finite transformations of the group (19.35)!

So it makes sense to distinguish only between those solutions (and their underlying subgroups), which *cannot* be transformed into each other by finite transformations of the (complete) group. But for multiparameter groups such as, for example, the 17-parameter group of the source-free Maxwell equations, even the determination of these inequivalent subgroups is a difficult task, which needs more knowledge of Lie groups than we have provided so far. The interested reader is therefore referred to the literature on this subject.

19.4 Exercises

1 Find all invariants and differential invariants (up to first order) of the symmetries (19.18). How are they related to the variables used when reducing the order of differential equation (19.20)?

2 Find the similarity variables of the general symmetry (16.66) and use it to reduce the sine–Gordon equation (16.65). How many symmetries does the resulting ordinary differential equation inherit?

3 Give the complete list of all one-variable solutions $u(v)$ of the three-dimensional potential equation (19.34) with respect to the group with generators (19.35).

20

Symmetries and the separability of partial differential equations

20.1 The problem

As already discussed in Section 18.1, one is often forced to give the solution to a set of partial differential equations in terms of functions each depending only on one variable. The approach we want to discuss here is that of the so-called separation of variables. A system $H_A = 0$ of partial differential equations is (completely) *separable* if in suitable coordinates (i.e., possibly only after a point transformation) it can be split into a set of ordinary differential equations by making an appropriate *ansatz* for the solutions. In the case of the N-dimensional wave equation in a curved spaced or in curvilinear coordinates,

$$\Box u = \frac{1}{\sqrt{|g|}}(g^{ab}u_{,b}\sqrt{|g|})_{,a} = 0 \tag{20.1}$$

(g_{ab} being the metric of this space, $g = \|g_{ab}\|$ its determinant, and g^{ab} the inverse of g_{ab}), this *ansatz* is

$$u = \overset{(1)}{u}(x^1)\overset{(2)}{u}(x^2)\cdots\overset{(N)}{u}(x^N), \tag{20.2}$$

and for the Hamilton–Jacobi equation of a freely moving particle in that space,

$$\tfrac{1}{2}g^{ab}W_{,a}W_{,b} - E = 0, \tag{20.3}$$

one starts with

$$W(x^i) = \overset{(1)}{W}(x^1) + \overset{(2)}{W}(x^2) + \cdots + \overset{(N)}{W}(x^N). \tag{20.4}$$

In a certain sense, the form of the *ansatz* (20.4) as a sum is a consequence of taking a product in (20.2): if we are interested in the rays of the wave field described by (20.1), then we would make an eikonal *ansatz*

$$u = u_0(x^n)e^{i\omega W(x^n)}. \tag{20.5}$$

Neglecting all terms except those with ω^2 will then lead from (20.1) to (20.3) with $E = 0$, and writing u as a product results in writing W as a sum.

Even for the most frequently discussed partial differential equations such as the wave equation and the Hamilton–Jacobi equation, the problem of finding all coordinates or spaces in which (20.2) and (20.4) do not lead to contradictions is a rather involved one. So we shall give only some more or less intuitive remarks on the main ideas and refer the reader to the vast literature on this field.

One point should be stressed beforehand. Whereas when using similarity variables we were led to rather special solutions, the importance of the separation of variables lies in the fact that the most general solution can be obtained this way – either, as for the wave equation, in the form of a superposition of special solutions or, as for the Hamilton–Jacobi equation, because we are interested only in solutions depending on a certain set of N parameters.

We shall now proceed as follows. In Section 20.2, we collect some hints that symmetries may be connected with the usual separation techniques of the four-dimensional wave equation. Section 20.3 gives some facts on the connection between the Hamilton–Jacobi equation and Hamilton's canonical equations and on first integrals of the latter that are in involution. In Section 20.4 the theorem is stated that shows the interrelations among symmetries, first integrals in involution, and separability of the Hamilton–Jacobi equation and the wave equation.

20.2 Some remarks on the usual separations of the wave equation

When separating the wave equation in Minkowski space,

$$\Box u = \left(\frac{\partial^2}{\partial x^2} + \frac{\partial^2}{\partial y^2} + \frac{\partial^2}{\partial z^2} - \frac{\partial^2}{\partial t^2}\right)u = 0, \tag{20.6}$$

in Cartesian coordinates, one obtains the particular solution (plane wave)

$$u = Ae^{ik_a x^a} = e^{i(k_1 x + k_2 y + k_3 z - \omega t)}, \qquad k_1^2 + k_2^2 + k_3^2 - \omega^2 = 0. \tag{20.7}$$

The general solution can be represented by a superposition of those plane waves as a Fourier integral.

When separating the same equation in spherical coordinates, one has the particular solution

$$u = A(\omega r)^{1/2} I_{n+1/2}(\omega r) P_n^m(\cos \vartheta) e^{im\varphi} e^{-i\omega t}, \tag{20.8}$$

where the $I_{n+1/2}$ and P_n^m are Bessel functions and Legendre functions, respectively, and again the general solution can be given as a sum over such particular solutions.

Many branches of analysis and mathematical physics rest on the fact that for a linear differential equation, such as the wave equation, the symmetries of the differential equation give rise to a linear representation of the symmetry group in the space of solutions and are reflected in special properties of the functions appearing in the separation (e.g., in the addition theorems of the Legendre functions).

Is there also a conceivable connection between the possibility of separating the wave equation in these two coordinate systems and the symmetries of the wave equation?

Among the symmetries of the wave equation, the most important ones in this connection are those forming the Poincaré group, generated by the four translations

$$\mathbf{X}_a = \frac{\partial}{\partial x^a} \tag{20.9}$$

and the six rotations [three rotations in (x, y, z)-space and three Lorentz transformations]

$$\mathbf{Y}_{ab} = -\mathbf{Y}_{ba} = \varepsilon_{ab}{}^{n}{}_{m} x^m \frac{\partial}{\partial x^n}, \qquad a, n, \ldots = 1, \ldots, 4 \tag{20.10}$$

(where ε_{abnm} is completely antisymmetric, and $\varepsilon_{1234} = -1$).

Clearly the Cartesian coordinates used in the separation (20.7) are related to the translational symmetries, the coordinate lines being the orbits of the four commuting generators $\mathbf{X}_a = \partial/\partial x^a$ and the functions $\overset{(1)}{u} = e^{ik_1 x}, \ldots$ being eigenfunctions of these generators. It is likewise clear that the spherical coordinates, and the coordinate t, used in (20.7) are related to the rotational symmetry with generators

$$\mathbf{X}_\beta = \varepsilon_{4\beta}{}^{\nu}{}_{\mu} x^\mu \frac{\partial}{\partial x^\nu}, \qquad \beta, \nu, \mu = 1, 2, 3, \tag{20.11}$$

and the time translation $\mathbf{X}_4 = \partial/\partial t$, although the explicit connection between the generators $\mathbf{X}_4$, $\mathbf{Y}_\beta$ and, for example, the coordinate ϑ is less obvious. Again the functions appearing in the solution (20.8) are eigenfunctions, not, however, of the four symmetries $\mathbf{X}_4$, $\mathbf{Y}_\beta$ but rather of $\mathbf{X}_4$, $\mathbf{Y} = \partial/\partial\varphi$ (which is one of the $\mathbf{Y}_\beta$ if written in spherical coordinates) and

$$\mathbf{Y}^\beta \mathbf{Y}_\beta = \frac{1}{\sin\vartheta}\frac{\partial}{\partial\vartheta}\sin\vartheta\frac{\partial}{\partial\vartheta} - \frac{1}{\sin^2\vartheta}\frac{\partial^2}{\partial\varphi^2} \tag{20.12}$$

(note that the three operators $\mathbf{X}_4$, $\mathbf{Y}$, and $\mathbf{Y}_\beta\mathbf{Y}^\beta$ commute!).

So these two examples of coordinate systems in which the wave equation separates indicate that there is a connection between separability and the existence of symmetries. But it is not yet clear how a systematic search for such coordinate systems should be carried out and whether the rest of the full symmetry group (e.g., the scaling symmetry with generator $\mathbf{X} = x^n\,\partial/\partial x^n$) could also be used for that purpose.

20.3 Hamilton's canonical equations and first integrals in involution

To make the connection between separability of the Hamilton–Jacobi equation and the existence of certain first integrals of geodesic motion understandable, we want to remind the reader of some well-known facts of classical analytical mechanics.

There are (at least) three different, but equivalent formulations of the equations of motion, namely, the Lagrangian, the Hamiltonian, and the Hamilton–Jacobi approach.

In the Lagrangian approach (used in Section 13.4), the equations of geodesic motion are written as the Euler–Lagrange equations

$$\frac{\mathrm{d}}{\mathrm{d}s}\frac{\partial L}{\partial \dot{x}^k} - \frac{\partial L}{\partial x^k} = 0, \qquad \dot{x}^k \equiv \frac{\mathrm{d}x^k}{\mathrm{d}s} \tag{20.13}$$

of the Lagrangian

$$L(x^k, \dot{x}^k) = \tfrac{1}{2} g_{ab}(x^n)\dot{x}^a\dot{x}^b. \tag{20.14}$$

If we introduce instead of the $\dot{x}^k$ and L new variables, namely, the generalized momenta p_k and the Hamiltonian H defined by

$$p_k = \frac{\partial L}{\partial \dot{x}^k}, \qquad H = -L + p_k\dot{x}^k, \tag{20.15}$$

then we arrive at the canonical equations

$$\dot{p}_k = -\frac{\partial H}{\partial x^k}, \qquad \dot{x}_k = \frac{\partial H}{\partial p_k}, \tag{20.16}$$

where the Hamiltonian is given by

$$H(x^k, p_k) = \tfrac{1}{2}g^{ab}(x^n)p_a p_b. \tag{20.17}$$

Transformations $\{x^k, p_k, H\} \to \{Q^k, P_k, H'\}$, which preserve the canonical equations (20.16), are called canonical transformations. Such transformations could be generated from a function $S(x^k, P_k, s)$ by

$$Q^k = \frac{\partial S}{\partial P_k}, \qquad p_k = \frac{\partial S}{\partial x^k}, \qquad H' = \frac{\partial S}{\partial s}. \tag{20.18}$$

Canonical transformations can be used to simplify the canonical equations. The most drastic simplification is to have a zero Hamiltonian H' because in that case the canonical equations (20.16) give

$$P_k = \text{const.}, \qquad Q^k = \text{const.} \tag{20.19}$$

The generating function $S(x^k, P_k, x)$, which would accomplish this transformation, would have to satisfy $H' = H + \partial S/\partial s = 0$; that is, because $p_k = \partial S/\partial x^k$, it has to satisfy the general Hamilton–Jacobi equation

$$H(x^k, \partial S/\partial x^k, s) + \partial S/\partial s = 0. \tag{20.20}$$

For the Hamiltonian (20.17), which does not depend explicitly on s, we can make the separation *ansatz*

$$S = W(x^k, P_k) - Es. \tag{20.21}$$

Equations (20.17) and (20.21) then imply the Hamilton–Jacobi equation

$$\tfrac{1}{2}g^{ab}(x^n)W_{,a}W_{,b} - E = 0 \tag{20.22}$$

in the form discussed in the preceding sections. The P_k do not really show up in this equation, but what we need is a solution W that genuinely depends on N constants P_k, where N is the dimension of the space (i.e., the number of the x^k).

Solving the N second order differential equations of geodesic motion (20.13) is equivalent to finding $2N$ first integrals φ^α. A special class of these first integrals is of particular importance in the Hamiltonian and, as we shall show later, also in the Hamilton–Jacobi approach. These are the first integrals in involution. They can best be characterized by using Poisson brackets.

By definition, the Poisson bracket of any two functions $F(x^k, p_k)$ and $G(x^k, p_k)$ is

$$\{F, G\} = \frac{\partial F}{\partial x^k}\frac{\partial G}{\partial p_k} - \frac{\partial F}{\partial p_k}\frac{\partial G}{\partial x^k}. \tag{20.23}$$

Poisson brackets can be used to characterize the independence and completeness of the system x^k, p_k of variables,

$$\{p_k, p_j\} = 0 = \{x^k, x^j\}, \qquad \{p_k, x^j\} = \delta^j_k, \tag{20.24}$$

and to reformulate the canonical equations (20.16) as

$$\{p_k, H\} = \dot{p}_k, \qquad \{x^k, H\} = \dot{x}^k. \tag{20.25}$$

We consider now first integrals $\varphi^\alpha(x^j, p_k)$ that are independent of s. Because of

$$\frac{d\varphi^\alpha}{ds} = \frac{\partial\varphi^\alpha}{\partial x^k}\dot{x}^k + \frac{\partial\varphi}{\partial p_k}\dot{p}_k = 0 \tag{20.26}$$

and the canonical equations, they satisfy

$$\{\varphi^\alpha, H\} = 0. \tag{20.27}$$

Since Poisson brackets fulfil the Jacobi identity

$$\{F, \{G, L\}\} + \{G, \{L, F\}\} + \{L, \{F, G\}\} = 0, \tag{20.28}$$

it follows that the Poisson bracket of any two first integrals φ^1 and φ^2 is again a first integral,

$$\{\varphi^1, \varphi^2\} = \varphi^3, \tag{20.29}$$

which, however, need not be functionally independent of φ^1 and φ^2. It may even happen that for a certain class of first integrals

$$\{\varphi^\alpha, \varphi^\beta\} = 0, \qquad \{\varphi^\alpha, H\} = 0 \tag{20.30}$$

holds. We then say that these first integrals (and H) are *in involution.*

In N-dimensional space ($k = 1, \ldots, N$), there are exactly N functionally independent first integrals in involution, including H: because of the third set of equations (20.24), there cannot be more than N, and the maximum number N is, for example, provided by the $P_k(x^n, p_n)$.

The complete symmetry between the first integrals φ^α and the Hamiltonian H in (20.30) shows that one may also think of one of the φ^α as being the Hamiltonian (of a different physical system, of course) and H as just a first integral.

By the way, a set of N first integrals in involution corresponds to a complete set of commuting observables in quantum mechanics.

20.4 Quadratic first integrals in involution and the separability of the Hamilton–Jacobi equation and the wave equation

First integrals φ^α are on an equal footing with the Hamiltonian $H = g^{ab}p_a p_b/2$ if they are quadratic in momenta p_a, that is, if they have the form

$$\varphi^\alpha = \overset{\alpha}{K}{}^{ab}(x^k)p_a p_b = \overset{\alpha}{K}_{ab}\dot{x}^a\dot{x}^b \tag{20.31}$$

(where we have used the relation $p_a = g_{ab}\dot{x}^b$ between momenta p_a and velocities $\dot{x}^b$).

First integrals of this type have already been discussed at some length in Section 13.4. The main results obtained there were that the $\overset{\alpha}{K}_{ab}$ have to be Killing tensors, that is, they must satisfy

$$\overset{\alpha}{K}_{ab;c} + \overset{\alpha}{K}_{bc;a} + \overset{\alpha}{K}_{ca;b} = 0, \tag{20.32}$$

and that besides the genuine Killing tensors there are trivial ones that are combinations of the metric tensor

$$K_{ab} = g_{ab} \tag{20.33}$$

and of products of Killing vectors such as

$$\overset{\alpha}{K}_{ab} = \overset{\alpha}{K}_a\overset{\alpha}{K}_b, \qquad \overset{\alpha}{K}_{a;b} + \overset{\alpha}{K}_{b;a} = 0. \tag{20.34}$$

The case (20.33) leads to the Hamiltonian H (which is of course also a quadratic first integral), whereas (20.34) gives first integrals

$$\varphi^\alpha = (\overset{\alpha}{K}{}^a p_a)^2, \tag{20.35}$$

which are squares of linear first integrals. Both can be used when constructing a set of functionally independent first integrals, but then products of two different Killing vectors $\overset{\alpha}{K}_a$ and $\overset{\beta}{K}_a$,

$$\overset{\lambda}{K}_{ab} = \overset{\alpha}{K}_a\overset{\beta}{K}_b + \overset{\alpha}{K}_b\overset{\beta}{K}_a, \tag{20.36}$$

need not be taken into account: they are functionally dependent on those built according to (20.35) from $\overset{\alpha}{K}_a$ and $\overset{\beta}{K}_a$.

When are two quadratic first integrals in involution? In terms of the corresponding Killing tensors, we obtain from the definition (20.23) of the Poisson bracket

$$\{\varphi^\alpha, \varphi^\beta\} = \{\overset{\alpha}{K}{}^{ab} p_a p_b, \overset{\beta}{K}{}^{nm} p_n p_m\}$$
$$= 2\overset{\alpha}{K}{}^{ab}{}_{,k} \overset{\beta}{K}{}^{kn} p_a p_b p_n - 2\overset{\alpha}{K}{}^{ak} p_a \overset{\beta}{K}{}^{nm}{}_{,k} p_n p_m = 0, \tag{20.37}$$

from which

$$\overset{\alpha}{K}{}^{k(b} \overset{\beta}{K}{}^{nm)}{}_{,k} - \overset{\alpha}{K}{}^{(nm}{}_{,k} \overset{\beta}{K}{}^{b)k} = 0 \tag{20.38}$$

follows (the parentheses around the indices b, n, m denote that the symmetric part has to be taken). If two Killing tensors $\overset{\alpha}{K}{}^{ab}$, $\overset{\beta}{K}{}^{nm}$ satisfy this condition, they are called *commuting*. The reason for this notion becomes transparent when we apply (20.38) to the special trivial Killing tensors $\overset{\alpha}{K}{}^{ab} = \overset{\alpha}{K}{}^{a} \overset{\alpha}{K}{}^{b}$. Equation (20.38) then reads

$$\overset{\alpha}{K}{}^{k} \overset{\beta}{K}{}^{n}{}_{,k} - \overset{\beta}{K}{}^{k} \overset{\alpha}{K}{}^{n}{}_{,k} = 0, \tag{20.39}$$

[best taken directly from (20.37)] and if we remember that Killing vectors $\overset{\alpha}{K}_a$ are related to the symmetry generators $\mathbf{X}$ of the Euler–Lagrange equations (compare Section 13.4) by

$$\overset{\alpha}{\mathbf{X}} = \overset{\alpha}{K}{}^{a} \frac{\partial}{\partial x^a}, \tag{20.40}$$

we see that (20.40) simply states that the two corresponding generators commute,

$$[\overset{\alpha}{\mathbf{X}}, \overset{\beta}{\mathbf{X}}] = 0. \tag{20.41}$$

We now have to hand the necessary tools to formulate the long-promised theorem on the separability. Here it is:

An orthogonal system of coordinates ($g_{ab} = 0$ for $a \neq b$; $a, b = 1, \ldots, N$), in which the Hamilton–Jacobi equation $g^{ab} W_{,a} W_{,b} = 2E$ is completely separable exists exactly if there are N independent commuting

21

Contact transformations and contact symmetries of partial differential equations, and how to use them

21.1 The general contact transformation and its infinitesimal generator

So far we have been concerned with point transformations of partial differential equations, that is, transformations in the space of variables (x^n, u^α), and with the symmetries a system of partial differential equations may have under such transformations. From the case of ordinary differential equations we know that it was possible to generalize these transformations to contact transformations by including first derivatives and to dynamical transformations by allowing for even higher derivatives.

Do similar generalizations make sense also for partial differential equations? The answer is yes, it makes sense, and most of the following chapters will be devoted to this problem. But as in the case of ordinary differential equations, a warning should be given to the reader: the more general the allowed transformations are, the more symmetries can exist – but the more difficult it will be to find and to exploit them.

The first step in generalizing point transformations is to allow for first derivatives, that is, to consider transformations

$$\tilde{x}^n = \tilde{x}^n(x^i, u^\beta, u^\beta{}_{,i}), \qquad \tilde{u}^\alpha = \tilde{u}^\alpha(x^i, u^\beta, u^\beta{}_{,i}),$$
$$\tilde{u}^\alpha{}_{,n} = \tilde{u}^\alpha{}_{,n}(x^i, u^\beta, u^\beta{}_{,i}) \tag{21.1}$$

in the space of variables $(x^i, u^\beta, u^\beta{}_{,i})$. We call them *contact transformations* if they satisfy the contact condition that the transformed derivatives $\tilde{u}^\alpha{}_{,n}$ are the extensions to the derivatives of the transformations of the x^n and u^α, that

Killing tensors K_{ab} that have a common system of orthogonal, hypersurface orthogonal eigenvectors. If in this system of coordinates the Ricci tensor R_{ab} [compare (16.68)] is diagonal too, then the wave equation $(g^{ab}u_{,a}\sqrt{|g|})_{,b}/\sqrt{|g|}=0$ is also completely separable. On the other hand, if in this system of coordinates a function $V(x^n)$ has the form $V=V(x^1)g^{11} + V(x^2)g^{22}+\cdots+V(x^N)g^{NN}$, then $g^{ab}W_{,a}W_{,b}+2V=2E$ is also completely separable.

This theorem shows that separability is intimately connected with those symmetries of the equations of geodesic motion that show up in the existence of special second order Killing tensors. Whether a Riemannian space admits such Killing tensors can be checked (in principle) in any given system of coordinates.

20.5 Exercises

1 Show that (20.38) – when applied to g_{ab} and any K_{ab} – gives the definition (20.32) of a Killing tensor.

2 Can it happen that H and the first integrals (20.35) are functionally dependent?

3 Which are the Killing tensors used in the standard separation (20.8) of the wave equation?

is, if

$$\tilde{u}^{\alpha}{}_{,n} = \frac{\partial \tilde{u}^{\alpha}}{\partial \tilde{x}^{n}} \tag{21.2}$$

holds. In terms of the infinitesimal generator

$$\mathbf{X} = \xi^{n}(x^{i}, u^{\beta}, u^{\beta}{}_{,i})\frac{\partial}{\partial x^{n}} + \eta^{\alpha}(x^{i}, u^{\beta}, u^{\beta}{}_{,i})\frac{\partial}{\partial u^{\alpha}} + \eta^{\alpha}_{n}(x^{i}, u^{\beta}, u^{\beta}{}_{,i})\frac{\partial}{\partial u^{\alpha}{}_{,n}}, \tag{21.3}$$

this condition means that the η^{α}_{i} are the extensions of ξ^{n} and η^{α}, which therefore have to obey

$$\eta^{\alpha}_{i} = \frac{\mathrm{D}\eta^{\alpha}}{\mathrm{D}x^{i}} - u^{\alpha}{}_{,n}\frac{\mathrm{D}\xi^{n}}{\mathrm{D}x^{i}}, \qquad \frac{\mathrm{D}}{\mathrm{D}x^{i}} = \frac{\partial}{\partial x^{i}} + u^{\beta}{}_{,i}\frac{\partial}{\partial u^{\beta}} + u^{\beta}{}_{,ik}\frac{\partial}{\partial u^{\beta}{}_{,k}}; \tag{21.4}$$

compare Section 15.1. In particular, the η^{α}_{i} must be independent of the second derivatives $u^{\beta}{}_{,ik}$, so that in view of (21.4)

$$\frac{\partial \eta^{\alpha}}{\partial u^{\beta}{}_{,k}} = u^{\alpha}{}_{,n}\frac{\partial \xi^{n}}{\partial u^{\beta}{}_{,k}} \tag{21.5}$$

follows.

To see whether transformations of this kind can exist at all, we evaluate the integrability conditions

$$\frac{\partial^{2}\eta^{\alpha}}{\partial u^{\gamma}{}_{,i}\,\partial u^{\beta}{}_{,k}} - \frac{\partial^{2}\eta^{\alpha}}{\partial u^{\beta}{}_{,k}\,\partial u^{\gamma}{}_{,i}} = 0 = \delta^{\alpha}_{\gamma}\frac{\partial \xi^{i}}{\partial u^{\beta}{}_{,k}} - \delta^{\alpha}_{\beta}\frac{\partial \xi^{k}}{\partial u^{\gamma}{}_{,i}} \tag{21.6}$$

of equation (21.5).

If there is more than one function u^{α},we can choose $\alpha \neq \gamma$, $\alpha = \beta$, in the right side of equation (21.6) and obtain

$$\frac{\partial \xi^{k}}{\partial u^{\gamma}{}_{,i}} = 0; \tag{21.7}$$

the functions ξ^{k} and also – because of (21.5) – the functions η^{α} do not depend on the derivatives: all contact transformations in more than one dependent variable are extended point transformations. This negative result is a severe limit on applying contact transformations (and symmetries) to partial differential equations.

If there is only one dependent variable u, then the integrability condition (21.6) is satisfied trivially, and the extension conditions (21.4) and (21.5) read

$$\eta_i = \eta_{,i} + \eta_{,u}u_{,i} - u_{,n}\xi^n{}_{,i} - u_{,n}u_{,i}\xi^n{}_{,u}, \tag{21.8}$$

$$\frac{\partial\eta}{\partial u_{,k}} = u_{,n}\frac{\partial\xi^n}{\partial u_{,k}}. \tag{21.9}$$

Once (21.9) is satisfied, we can use (21.8) to calculate the η_i. Functions η, ξ^n satisfying (21.9) exist. They are best given in terms of a generating function $\Omega(x^n, u, u_{,n}) = u_{,n}\xi^n - \eta$. We leave the proof to the reader and summarize the result as follows:

Every infinitesimal contact transformation

$$\mathbf{X} = \xi^n(x^i, u, u_{,i})\frac{\partial}{\partial x^n} + \eta(x^i, u, u_{,i})\frac{\partial}{\partial u} + \eta_n(x^i, u, u_{,i})\frac{\partial}{\partial u_{,n}} \tag{21.10}$$

can be obtained from a generating function $\Omega(x^i, u, u_{,i})$ *according to*

$$\xi^n = \frac{\partial\Omega}{\partial u_{,n}}, \qquad \eta = u_{,i}\frac{\partial\Omega}{\partial u_{,i}} - \Omega, \qquad \eta_i = -\left(u_{,i}\frac{\partial\Omega}{\partial u} + \frac{\partial\Omega}{\partial x^i}\right). \tag{21.11}$$

If Ω is linear in the derivatives $u_{,i}$, then the transformation is an extended point transformation.

21.2 Contact symmetries of partial differential equations and how to find them

A contact transformation with generator $\mathbf{X}$ will be called a symmetry of a partial differential equation $H = 0$ if

$$\mathbf{X}H(x^i, u, u_{,i}, \ldots) \equiv 0 \qquad (\mathrm{mod}\, H = 0) \tag{21.12}$$

holds. The difference from a Lie point symmetry is that ξ^n, η are in general also functions of the derivatives and that no genuine contact symmetry for more than one dependent variable exists.

The connections between *first order partial differential equations* and contact transformations and symmetries have been extensively discussed in the old literature, and we do not want to go into the details here. We note that for the Hamilton–Jacobi equation

$$H = \hat{H}(x^i, u_{,i}) - E = 0 \tag{21.13}$$

the symmetry condition $\mathbf{X}H = 0$ can in view of Exercise 21.2 be written as

$$\{\Omega, \hat{H}\} = 0 \qquad (\mathrm{mod}\, \hat{H} = E). \tag{21.14}$$

So finding all contact symmetries essentially means finding all first integrals; compare Section 19.3. And we further note that any two first order differential equations with the same number of independent variables can be transformed into each other by a contact transformation, which again indicates that finding all contact symmetries means finding the general solution.

For a *second order partial differential equation* we start with

$$\mathbf{X} = \frac{\partial \Omega}{\partial u_{,i}} \frac{\partial}{\partial x^i} + \left(u_{,i} \frac{\partial \Omega}{\partial u_{,i}} - \Omega \right) \frac{\partial}{\partial u} - \left(\frac{\partial \Omega}{\partial u} u_{,i} + \Omega_{,i} \right) \frac{\partial}{\partial u_{,i}} + \eta_{ik} \frac{\partial}{\partial u_{,ik}}, \tag{21.15}$$

$$\begin{aligned} \eta_{ik} = & -u_{,in} u_{,mk} \frac{\partial^2 \Omega}{\partial u_{,n} \partial u_{,m}} - (u_{,i} u_{,nk} + u_{,k} u_{,in}) \frac{\partial^2 \Omega}{\partial u\, \partial u_{,n}} \\ & - u_{,ik} \frac{\partial \Omega}{\partial u} - u_{,nk} \frac{\partial^2 \Omega}{\partial x^i \partial u_{,n}} - u_{,in} \frac{\partial^2 \Omega}{\partial x^k \partial u_{,n}} - u_{,i} \frac{\partial^2 \Omega}{\partial x^k \partial u} \\ & - u_{,k} \frac{\partial^2 \Omega}{\partial x^i \partial u} - u_{,i} u_{,k} \frac{\partial^2 \Omega}{\partial u^2} - \frac{\partial^2 \Omega}{\partial x^i \partial x^k} \end{aligned} \tag{21.16}$$

and determine the function $\Omega(x^i, u, u_{,i})$ from the symmetry condition (21.12). Although this is a straightforward task, the differential equations for Ω that arise from equating to zero the coefficients of, for example, the different powers of the second derivatives will in general be rather difficult to treat.

As an easy example we take the heat conduction equation

$$u_{,xx} - u_{,t} = 0. \tag{21.17}$$

If for brevity we write

$$u_{,x} = p, \qquad u_{,t} = q \tag{21.18}$$

and use the field equation to eliminate $u_{,xx}$, then we have from the definition (21.16) of the η_{ik}

$$\eta_{xx} = -u_{,xy}^2 \Omega_{,qq} - 2qu_{,xy} \Omega_{,pq} - 2pu_{,xy} \Omega_{,uq} - 2u_{,xy} \Omega_{,xq} + \cdots, \tag{21.19}$$

where the ellipses indicate terms without second derivatives. From the symmetry condition

$$\eta_{xx} - \eta_t \equiv 0, \tag{21.20}$$

one thus immediately obtains, taking into account only the terms with $u_{,xy}$,

$$\Omega = A(t)q + B(x, t, p). \tag{21.21}$$

Inserting now η_{xx} and η_t in full, one sees that Ω is linear in both q and p, that is, in $u_{,x}$ and $u_{,t}$, which characterizes a point transformation. So all contact symmetries of the heat conduction equation are point symmetries (except possibly those transformations that are not members of a one-parameter group).

Two other examples of partial differential equations that *do* admit a contact symmetry, together with the generating functions Ω of one of the symmetries they have, are

$$u_{,xx} + Bu_{,xy} + u_{,yy} = 0, \qquad B = \text{const.},$$
$$\Omega = \tfrac{1}{2}(u_{,x}^2 - u_{,y}^2) \tag{21.22}$$

and

$$u_{,xx}u_{,yy} - u_{,xy}^2 - 1 = 0, \qquad \Omega = \tfrac{1}{2}(xu_{,x}^2 - xy^2) - yu_{,x}u_{,y}. \tag{21.23}$$

21.3 Remarks on how to use contact symmetries for reduction of variables

Contact symmetries have not been widely used for integration procedures (except for first order differential equations where such ideas were frequently used in the older literature). This is most probably due to the fact that genuine contact symmetries do not often exist and are difficult to find and systems of differential equations with more than one dependent variable do not admit them at all. We shall therefore make only a few remarks on how to use them and concentrate on the possibility of reducing the number of variables by means of a known contact symmetry.

Suppose a differential equation $H = 0$ admits a contact symmetry $\mathbf{X}$ with the generating function Ω. We then consider the first order differential equation

$$\Omega(x^n, u, u_{,n}) = 0. \tag{21.24}$$

If we apply the operator $\mathbf{X}$ – as given by (21.15) – to this equation, we obtain

$$\mathbf{X}\Omega = \frac{\partial\Omega}{\partial u_{,n}}\frac{\partial\Omega}{\partial x^n} + \left(u_{,i}\frac{\partial\Omega}{\partial u_{,i}} - \Omega\right)\frac{\partial\Omega}{\partial u} - \left(u_{,n}\frac{\partial\Omega}{\partial u} + \frac{\partial\Omega}{\partial x^n}\right)\frac{\partial\Omega}{\partial u_{,n}}$$

$$= -\Omega\frac{\partial \Omega}{\partial u} = 0. \qquad (21.25)$$

That is, the operator **X** leaves not only $H = 0$ but also $\Omega = 0$ invariant; it is a contact symmetry of these *two* differential equations:

$$\mathbf{X}H = 0 \quad (\operatorname{mod} H = 0), \qquad \mathbf{X}\Omega = 0 \quad (\operatorname{mod} \Omega = 0). \qquad (21.26)$$

This set of equations implies that $\Omega = 0$ and $H = 0$ have a common set of solutions, which are exactly the counterpart of the similarity solutions of a Lie point symmetry. To see this, imagine we have transformed the operator **X** to its normal form

$$\tilde{\mathbf{X}} = \frac{\partial}{\partial \psi}, \qquad \tilde{\Omega} = \tilde{u}_{,\psi} \qquad (21.27)$$

by a coordinate (contact!) transformation in $(x^n, u, u_{,n})$-space. Then $\tilde{\Omega} = 0$ says that this common set is characterized by $\tilde{u}_{,\psi} = 0$, and, because of $\partial \tilde{H}/\partial \psi = 0$, this does not contradict $\tilde{H} = 0$; compare the discussion in connection with similarity variables in Sections 18.1 and 18.2.

Because of the relation (21.11) between Ω, η, and ξ^n, we can also write the differential equation $\Omega = 0$ as

$$\eta - \xi^n u_{,n} = 0, \qquad (21.28)$$

which differs from equation (18.7) characterizing similarity solutions of Lie point symmetries only in that here η and ξ^n can depend on the derivatives $u_{,i}$, which they must not in (18.7), and that here only one dependent variable u is admitted.

To find the common set of solutions, one can (in principle) take $\Omega = 0$ and solve it with respect to, for example, $u_{,1}$. This expression can then be used to eliminate $u_{,1}, u_{,1n}, u_{,1nm}, \ldots$ from $H = 0$. By that process $H = 0$ becomes a differential equation in one less variable, since x_1 – if it appears – plays only the role of a parameter. If we have found the solution to this differential equation, we must subject it of course to $\Omega = 0$.

As an example, we take the potential equation and its symmetry (21.23);

$$H = u_{,xx} + u_{,yy} = 0, \qquad \Omega = \tfrac{1}{2}(u_{,x}^2 - u_{,y}^2) = 0. \qquad (21.29)$$

From $\Omega = 0$ we have

$$u_{,x} = \pm u_{,y}, \qquad u_{,xy} = \pm u_{,yy}, \qquad u_{,xx} = \pm u_{,xy} = u_{,yy}, \qquad (21.30)$$

so that $H = 0$ can be reduced to

$$H \equiv 2u_{,yy} = 0. \tag{21.31}$$

Its solution $u = A(x)y + B(x)$ must be subjected to $\Omega = 0$, so that

$$u = a(x \pm y) + b, \qquad a, b = \text{const.},$$

is the desired similarity solution.

21.4 Exercises

1 Show that for every contact transformation a generating function Ω exists that satisfies (21.11) and that for every function Ω the equations (21.11) define a contact transformation.

2 Suppose that the generating function Ω does not depend on u, and write p_i instead of $u_{,i}$. Show that the action of $\mathbf{X}$ on any function $\varphi(x^i, p_i)$ equals the Poisson bracket of φ and Ω, $\mathbf{X}\varphi = \{\varphi, \Omega\}$.

3 Show that the differential equations (21.22) and (21.23) admit the respective contact symmetries.

4 Show that all contact symmetries of the Liouville equation $u_{,xt} - e^u = 0$ are point symmetries.

5 Assume that a differential equation admits several contact symmetries $\mathbf{X}_A$ with generating functions Ω_A. Use the definition $\Omega = u_{,i}\xi^i - \eta$ to determine the generating function Ω_{AB} corresponding to the commutator $[\mathbf{X}_A, \mathbf{X}_B]$ and use Ω_{AB} to show that $\mathbf{X}_A\Omega_B = 0$ (mod $\Omega_C = 0$) holds. How can the contact symmetries be used for a multiple reduction of the number of variables in the given differential equation?

6 Show that the finite contact transformation $\tilde{u}_{,x} = x$, $\tilde{u}_{,y} = u_{,y}$, $\tilde{x} = -u_{,x}$, $\tilde{y} = y$, and $\tilde{u} = u - xu_{,x}$ transforms (21.22) into (21.23).

22

Differential equations and symmetries in the language of forms

22.1 Vectors and forms

Differential forms offer an alternative, more modern way of formulating systems of ordinary or partial differential equations and of getting insight into the structure of their point and contact symmetries. Since they are not so widely known among physicists, we have not used them so far, and we do not intend to give a detailed presentation of this part of the theory of differential manifolds. Rather we prefer a very pragmatic approach, starting with some basic definitions and then concentrating on the rules for dealing with forms and their exterior derivatives.

The basic objects we have to deal with are vectors (denoted by capital boldface latin letters such as $\mathbf{X}, \mathbf{A}$) and forms (denoted by small boldface greek letters such as $\boldsymbol{\sigma}$, $\boldsymbol{\omega}$) which exist in an n-dimensional differential manifold (coordinates x^i).

A vector $\mathbf{V}$ is a linear differential operator that at each point maps a differentiable (real-valued) function $f(x^n)$ into a (real) number. In a coordinate representation a vector $\mathbf{V}$ is given by

$$\mathbf{V} = v^a(x^i)\frac{\partial}{\partial x^a}, \tag{22.1}$$

where the functions $v^a(x^i)$ are the components of the vector $\mathbf{V}$. Examples of vectors already encountered are the operator $\mathbf{A}$ (the equivalent of an ordinary differential equation) and the symmetry generators $\mathbf{X}$.

By definition, a 1-form (Pfaffian form) $\boldsymbol{\sigma}$ maps a vector $\mathbf{V}$ into a (real) function, the *contraction*, denoted by $\langle \boldsymbol{\sigma}, \mathbf{V} \rangle$ or $\mathbf{V} \lrcorner \boldsymbol{\sigma}$. In a coordinate representation, a 1-form is given by

$$\boldsymbol{\sigma} = \sigma_a(x^i)\,\mathrm{d}x^a, \tag{22.2}$$

σ_a being the components of the form, and the contraction can be expressed in terms of the components v^a and σ_a as

$$\langle \boldsymbol{\sigma}, \mathbf{V} \rangle = \mathbf{V} \lrcorner \boldsymbol{\sigma} = v^a \sigma_a. \tag{22.3}$$

In an n-dimensional differential manifold, there exist exactly n linearly independent 1-forms. A basic set of 1-forms is, for example, provided by the $\mathrm{d}x^i$.

Since both functions f and 1-forms $\boldsymbol{\sigma}$ lead to real functions if associated with a vector $\mathbf{V}$, there exists a 1-form $\boldsymbol{\omega}$ derived from f, the differential $\mathrm{d}f$,

$$\boldsymbol{\omega} = \mathrm{d}f = f_{,a}\,\mathrm{d}x^a \tag{22.4}$$

so that

$$\mathbf{V}f = \mathbf{V} \lrcorner \mathrm{d}f. \tag{22.5}$$

By means of the *exterior product*, 1-forms can be used to construct higher forms. The exterior product of two 1-forms $\boldsymbol{\sigma}$ and $\boldsymbol{\omega}$, denoted by $\boldsymbol{\sigma} \wedge \boldsymbol{\omega}$, is linear in each variable and antisymmetric, that is, it vanishes if $\boldsymbol{\sigma}$ and $\boldsymbol{\omega}$ are linearly dependent. Its coordinate representation is

$$\boldsymbol{\sigma} \wedge \boldsymbol{\omega} = \sigma_a \omega_b\,\mathrm{d}x^a \wedge \mathrm{d}x^b = \tfrac{1}{2}(\sigma_a\omega_b - \sigma_b\omega_a)\mathrm{d}x^a \wedge \mathrm{d}x^b, \tag{22.6}$$

and the still ambiguous overall factor can be fixed by demanding

$$\left(\frac{\partial}{\partial x^i}\right) \lrcorner (\mathrm{d}x^a \wedge \mathrm{d}x^b) = \delta^a_i\,\mathrm{d}x^b - \delta^b_i\,\mathrm{d}x^a. \tag{22.7}$$

If we want to emphasize that $\boldsymbol{\sigma}$ and $\boldsymbol{\omega}$ are 1-forms and their exterior product is a 2-form, we shall denote that by, for example,

$$\underset{(1)}{\boldsymbol{\sigma}} \wedge \underset{(1)}{\boldsymbol{\omega}} = \underset{(2)}{\boldsymbol{\alpha}}. \tag{22.8}$$

The most general 2-form is a linear superposition of the $\binom{n}{2}$ 2-forms $\mathrm{d}x^a \wedge \mathrm{d}x^b$,

$$\underset{(2)}{\boldsymbol{\alpha}} = \alpha_{ab}\,\mathrm{d}x^a \wedge \mathrm{d}x^b, \qquad \alpha_{ab} = -\alpha_{ba}. \tag{22.9}$$

By an obvious generalization one can introduce p-forms as the exterior product of p 1-forms. The general p-form $\underset{(p)}{\boldsymbol{\alpha}}$ is then given by

$$\underset{(p)}{\boldsymbol{\alpha}} = \alpha_{a_1 a_2 \cdots a_p} \mathrm{d}x^{a_1} \wedge \mathrm{d}x^{a_2} \wedge \cdots \wedge \mathrm{d}x^{a_p}, \tag{22.10}$$

where the coefficients $\alpha_{a_1 a_2 \cdots a_p}$ are totally antisymmetric,

$$\alpha_{a_1 a_2 \cdots a_p} = \alpha_{[a_1 a_2 \cdots a_p]}. \tag{22.11}$$

The exterior product of a p-form $\underset{(p)}{\boldsymbol{\alpha}}$ (with coefficients $\alpha_{a_1 \cdots a_p}$) and a q-form $\underset{(q)}{\boldsymbol{\beta}}$ (with coefficients $\beta_{b_1 \cdots b_q}$) is a $(p+q)$-form with coefficients

$$(\underset{(p)}{\boldsymbol{\alpha}} \wedge \underset{(q)}{\boldsymbol{\beta}})_{a_1 \cdots a_p b_1 \cdots b_q} = \alpha_{[a_1 \cdots a_p} \beta_{b_1 \cdots b_q]}, \tag{22.12}$$

which satisfies the commutation rule

$$\underset{(p)}{\boldsymbol{\alpha}} \wedge \underset{(q)}{\boldsymbol{\beta}} = (-1)^{pq} \underset{(q)}{\boldsymbol{\beta}} \wedge \underset{(p)}{\boldsymbol{\alpha}}. \tag{22.13}$$

In analogy with the contraction (22.3) of a vector $\mathbf{V}$ and a 1-form $\boldsymbol{\sigma}$, which is a function, the contraction of a vector $\mathbf{V}$ and a p-form $\underset{(p)}{\boldsymbol{\alpha}}$ gives a $(p-1)$-form $\underset{(p-1)}{\boldsymbol{\beta}}$,

$$\mathbf{V} \lrcorner \underset{(p)}{\boldsymbol{\alpha}} = \underset{(p-1)}{\boldsymbol{\beta}}. \tag{22.14}$$

In a component notation, the contraction is defined by

$$\begin{aligned} \mathbf{V} \lrcorner \underset{(p)}{\boldsymbol{\alpha}} &= \left(v^b \frac{\partial}{\partial x^b} \right) \lrcorner (\alpha_{[a_1 a_2 \cdots a_p]} \mathrm{d}x^{a_1} \wedge \mathrm{d}x^{a_2} \wedge \cdots \wedge \mathrm{d}x^{a_p}) \\ &= p! v^b \alpha_{[b a_2 a_3 \cdots a_p]} \mathrm{d}x^{a_2} \wedge \mathrm{d}x^{a_3} \wedge \cdots \wedge \mathrm{d}x^{a_p}, \end{aligned} \tag{22.15}$$

from which the components of $\underset{(p-1)}{\boldsymbol{\beta}}$ can easily be read off.

Two immediate consequences of the definition (22.15) are the rules

$$\mathbf{V} \lrcorner (\boldsymbol{\sigma} + \boldsymbol{\tau}) = \mathbf{V} \lrcorner \boldsymbol{\sigma} + \mathbf{V} \lrcorner \boldsymbol{\tau} \tag{22.16}$$

and

$$\mathbf{V} \lrcorner (\underset{(p)}{\boldsymbol{\alpha}} \wedge \underset{(q)}{\boldsymbol{\sigma}}) = (\mathbf{V} \lrcorner \underset{(p)}{\boldsymbol{\alpha}}) \wedge \underset{(q)}{\boldsymbol{\sigma}} + (-1)^p \underset{(p)}{\boldsymbol{\alpha}} \wedge (V \lrcorner \underset{(q)}{\boldsymbol{\sigma}}). \tag{22.17}$$

22.2 Exterior derivatives and Lie derivatives

The *exterior derivative* operator is a linear differential operator (denoted by d) that can be applied to forms and generates a $(p+1)$-form from a p-form. For a function f, the exterior derivative coincides with the ordinary derivative,

$$\mathrm{d}f = \frac{\partial f}{\partial x^a}\mathrm{d}x^a = \boldsymbol{\omega}; \tag{22.18}$$

it generates a 1-form ω from the 0-form (ordinary function) f. For arbitrary forms it satisfies

$$\mathrm{d}(\mathrm{d}\boldsymbol{\alpha}) = 0, \tag{22.19}$$

$$\begin{aligned}\mathrm{d}(\underset{(p)}{\boldsymbol{\alpha}} \wedge \underset{(q)}{\boldsymbol{\beta}}) &= (\mathrm{d}\underset{(p)}{\boldsymbol{\alpha}}) \wedge \underset{(q)}{\boldsymbol{\beta}} + (-1)^p \underset{(p)}{\boldsymbol{\alpha}} \wedge (\mathrm{d}\underset{(q)}{\boldsymbol{\beta}}),\\ \mathrm{d}(f\boldsymbol{\alpha}) &= \mathrm{d}f \wedge \boldsymbol{\alpha} + f\,\mathrm{d}\boldsymbol{\alpha},\end{aligned} \tag{22.20}$$

so that from the coordinate representation (22.10) of a p-form $\underset{(p)}{\boldsymbol{\alpha}}$, we obtain

$$\mathrm{d}\underset{(p)}{\boldsymbol{\alpha}} = \alpha_{a_1\cdots a_p,b}\,\mathrm{d}x^b \wedge \mathrm{d}x^{a_1} \wedge \cdots \wedge \mathrm{d}x^{a_p}. \tag{22.21}$$

As equation (22.19) shows, the repeated application of the exterior derivation yields zero. Conversely it can be shown that if $\boldsymbol{\alpha}$ is a p-form $(p \geqslant 1)$ such that $\mathrm{d}\boldsymbol{\alpha} = 0$, then there locally exists a $(p-1)$-form $\boldsymbol{\beta}$ such that $\boldsymbol{\alpha} = \mathrm{d}\boldsymbol{\beta}$ (Poincaré lemma).

The *Lie derivative* operator (denoted by $\mathscr{L}_{\mathbf{V}}$) is a linear differential operator associated with a vector field $\mathbf{V} = v^a\,\partial/\partial x^a$. It can be applied to both vectors and forms.

The Lie derivative (with respect to $\mathbf{V}$) of a vector $\mathbf{A}$ is the commutator of the vectors,

$$\mathscr{L}_{\mathbf{V}}\mathbf{A} = [\mathbf{V}, \mathbf{A}] = -\mathscr{L}_{\mathbf{A}}\mathbf{V}, \tag{22.22}$$

or in a coordinate representation

$$\mathscr{L}_{\mathbf{V}}\left(a^i\frac{\partial}{\partial x^i}\right) = (v^k a^i{}_{,k} - a^k v^i{}_{,k})\frac{\partial}{\partial x^i}, \tag{22.23}$$

that is, the Lie derivative of a vector is a vector.

The Lie derivative of a p-form is a p-form. For a function (or 0-form) f

the Lie derivative is simply the directional derivative with respect to $\mathbf{V}$,

$$\mathscr{L}_{\mathbf{V}} f = \mathbf{V} \lrcorner \mathrm{d}f = v^a f_{,a}, \tag{22.24}$$

and for a p-form $\boldsymbol{\alpha}$ (with $p > 0$) it can be given in terms of the exterior derivative and the contraction as

$$\mathscr{L}_{\mathbf{V}} \boldsymbol{\alpha} = \mathbf{V} \lrcorner \mathrm{d}\boldsymbol{\alpha} + \mathrm{d}(\mathbf{V} \lrcorner \boldsymbol{\alpha}). \tag{22.25}$$

Lie differentiation and exterior differentiation commute,

$$\mathscr{L}_{\mathbf{V}} \mathrm{d}\boldsymbol{\alpha} = \mathrm{d}\mathscr{L}_{\mathbf{V}} \boldsymbol{\alpha}. \tag{22.26}$$

The Lie derivatives of an exterior product and of a contraction are given by

$$\mathscr{L}_{\mathbf{V}}(\boldsymbol{\alpha} \wedge \boldsymbol{\beta}) = (\mathscr{L}_{\mathbf{V}} \boldsymbol{\alpha}) \wedge \boldsymbol{\beta} + \boldsymbol{\alpha} \wedge (\mathscr{L}_{\mathbf{V}} \boldsymbol{\beta}) \tag{22.27}$$

and

$$\mathscr{L}_{\mathbf{V}}(\mathbf{W} \lrcorner \boldsymbol{\alpha}) = [\mathbf{V}, \mathbf{W}] \lrcorner \boldsymbol{\alpha} + \mathbf{W} \lrcorner (\mathscr{L}_{\mathbf{V}} \boldsymbol{\alpha}). \tag{22.28}$$

22.3 Differential equations in the language of forms

To write a given differential equation (or system of differential equations) in the language of forms, the typical strategy is as follows.

First one writes the differential equation(s) as a set of quasi-linear first order differential equations by introducing the necessary number of first, second, and so on, derivatives as additional variables (additional to the originally given independent and dependent variables). This will be possible only if the original differential equation is linear in the highest derivatives.

In this enlarged set of variables, the above-mentioned system of quasi-linear first order differential equations essentially says that not all of these variables are independent of each other and that, therefore, (linear) relations between their differentials will exist. These relations can be written in terms of forms, and as a second step we must find a set of forms $\boldsymbol{\alpha}_i$ such that $\boldsymbol{\alpha}_i = 0$ is equivalent to the system of first order differential equations, that is, from $\boldsymbol{\alpha}_i = 0$ one recovers the original differential equation as well as the relations that ensure that some of the variables are in fact derivatives of the others. This set of forms should be closed, that is, if $\boldsymbol{\alpha}_i = 0$, then $\mathrm{d}\boldsymbol{\alpha}_i = 0$ should hold (as a consequence of being a linear combination of the $\boldsymbol{\alpha}_i$): in addition to the set of quasi-linear differential equations, their integrability conditions should also be satisfied.

To see how this procedure works, we consider two simple examples.

An ordinary second order differential equation

$$y'' = \omega(x, y, y') \tag{22.29}$$

is equivalent to the first order system

$$\frac{\mathrm{d}y}{\mathrm{d}x} = p, \qquad \frac{\mathrm{d}p}{\mathrm{d}x} = \omega(x, y, p), \tag{22.30}$$

that is, the enlarged set of variables is (x, y, p). The equations (22.30) show that there are two relations

$$\mathrm{d}y = p\,\mathrm{d}x, \qquad \mathrm{d}p = \omega\,\mathrm{d}x \tag{22.31}$$

between the differentials, and from them the forms $\boldsymbol{\alpha}_i$ are easily found to be

$$\boldsymbol{\alpha}_1 = \mathrm{d}y - p\,\mathrm{d}x, \qquad \boldsymbol{\alpha}_2 = \mathrm{d}p - \omega\,\mathrm{d}x. \tag{22.32}$$

The equations $\boldsymbol{\alpha}_i = 0$ are completely equivalent to the first order system (22.30) and thus to the original differential equation $y'' = \omega$. The set (22.32) of 1-forms is already closed, since because of

$$\begin{aligned} \mathrm{d}\boldsymbol{\alpha}_1 &= -\,\mathrm{d}p \wedge \mathrm{d}x = \mathrm{d}x \wedge \boldsymbol{\alpha}_2, \\ \mathrm{d}\boldsymbol{\alpha}_2 &= -\,\mathrm{d}\omega \wedge \mathrm{d}x = -\,\omega_{,y}\,\mathrm{d}y \wedge \mathrm{d}x - \omega_{,p}\,\mathrm{d}p \wedge \mathrm{d}x \\ &= \omega_{,y}\,\mathrm{d}x \wedge \boldsymbol{\alpha}_1 + \omega_{,p}\,\mathrm{d}x \wedge \boldsymbol{\alpha}_2, \end{aligned} \tag{22.33}$$

the 2-forms $\mathrm{d}\boldsymbol{\alpha}_i$ are linear in the $\boldsymbol{\alpha}_i$ and vanish if $\boldsymbol{\alpha}_i = 0$.

As an example of a partial differential equation, we take the heat conduction equation

$$u_{,xx} - u_{,t} = 0. \tag{22.34}$$

We introduce the first derivatives of u as new variables by

$$p = u_{,x}, \qquad q = u_{,t} \tag{22.35}$$

and associate the 1-form

$$\boldsymbol{\beta}_1 = \mathrm{d}u - p\,\mathrm{d}x - q\,\mathrm{d}t \tag{22.36}$$

with this first order system. In variables $\{x, t, u, p, q\}$, the heat conduction equation reads $\partial p/\partial x = q$, and this relation is equivalent to the 2-form

$$\boldsymbol{\beta}_2 = \mathrm{d}p \wedge \mathrm{d}t - q\,\mathrm{d}x \wedge \mathrm{d}t \tag{22.37}$$

in that it follows from $\boldsymbol{\beta}_2 = 0$ if we think of p and q as functions of x and t.

The system $\boldsymbol{\beta}_1, \boldsymbol{\beta}_2$ is not yet closed with respect to exterior differentiation. We have to add

$$\boldsymbol{\beta}_3 \equiv \mathrm{d}\boldsymbol{\beta}_1 = -\mathrm{d}p \wedge \mathrm{d}x - \mathrm{d}q \wedge \mathrm{d}t \tag{22.38}$$

since the right side cannot be expressed as a form linear in $\boldsymbol{\beta}_1$ and $\boldsymbol{\beta}_2$. The form $\boldsymbol{\beta}_3$ incorporates the integrability condition $\partial p/\partial t = u_{,xt} = \partial q/\partial x$, which follows from $\boldsymbol{\beta}_3 = 0$. The form $\mathrm{d}\boldsymbol{\beta}_2$ need not be added, as because of

$$\mathrm{d}\boldsymbol{\beta}_2 = -\mathrm{d}q \wedge \mathrm{d}x \wedge \mathrm{d}t = \mathrm{d}x \wedge \boldsymbol{\beta}_3, \tag{22.39}$$

it can be given in terms of $\boldsymbol{\beta}_3$.

In general, for a set of second order partial differential equations in N independent and M dependent variables, we expect to obtain M 1-forms from the introduction of the MN first derivatives as additional new variables, M 2-forms ensuring the integrability of the above first order set, and a set of N-forms from the original set of differential equations.

22.4 Symmetries of differential equations in the language of forms

So far symmetries have been always given in terms of an infinitesimal generator **X**, which, if properly extended, was a linear differential operator in the space of variables and derivatives that transformed as a vector in that space. So we know already that in the language of vectors and forms a symmetry is associated with a vector **X**. Of the two differential operators that naturally arise in the ring of forms, the exterior derivative and the Lie derivative, only the Lie derivative is based on a vector and will consequently be our candidate for formulating the symmetry conditions.

As discussed at some length in Sections 3.2 and 10.1, an ordinary differential equation (or a set of ordinary differential equations) can always be replaced by a first order linear partial differential equation

$$\mathbf{A}\varphi = a^k(x^i)\frac{\partial}{\partial x^k}\varphi = 0, \tag{22.40}$$

where the variables x^k are the independent variable, the dependent variables, and their derivatives up to a certain order. A symmetry generator **X** then has to satisfy

$$[\mathbf{X}, \mathbf{A}] = \lambda(x^i)\mathbf{A}. \tag{22.41}$$

Because of (22.22), this condition can be written as

$$\mathscr{L}_{\mathbf{X}}\mathbf{A} = \lambda\mathbf{A}. \tag{22.42}$$

This confirms that symmetries are to be expressed by means of the Lie derivative.

Symmetries of differential equations have always been defined by saying that the application of the symmetry operator to the differential equations (say, $H = 0$) should yield zero if the differential equations are properly taken into account. To translate this into the language of forms, an equivalent statement seems to be that the Lie derivative of all forms $\boldsymbol{\alpha}_i$ that represent the given differential equation(s) should vanish if $\boldsymbol{\alpha}_i = 0$ is used. The condition that $\mathscr{L}_{\mathbf{X}}\boldsymbol{\alpha}_i$ vanishes if $\boldsymbol{\alpha}_k = 0$ can also be formulated as saying that $\mathscr{L}_{\mathbf{X}} \underset{(p)}{\boldsymbol{\alpha}_i}$ must be forms linear in those $\underset{(q)}{\boldsymbol{\alpha}_k}$ for which $q \leqslant p$,

$$\mathscr{L}_{\mathbf{X}}\boldsymbol{\alpha}_i = \boldsymbol{\lambda}_i^k \wedge \boldsymbol{\alpha}_k, \tag{22.43}$$

where the $\boldsymbol{\lambda}_i^k$ are forms (if some or all of them are 0-forms, i.e., functions, we of course must write $\lambda_i^k \boldsymbol{\alpha}_k$ on the right side). We take (22.43) as the appropriate symmetry condition and say that the differential equations represented by the set of forms $\boldsymbol{\alpha}_i$ admit a symmetry with generator $\mathbf{X}$ exactly if (22.43) holds with some suitably chosen forms $\boldsymbol{\lambda}_i^k$.

To show that this definition coincides with the definition of a symmetry used so far, we consider two examples.

The *second order ordinary differential equation* $y'' = \omega(x, y, y')$ is equivalent to the system (22.32) of two 1-forms, that is, to

$$\boldsymbol{\alpha}_1 = \mathrm{d}y - p\,\mathrm{d}x, \qquad \boldsymbol{\alpha}_2 = \mathrm{d}p - \omega(x, y, p)\mathrm{d}x. \tag{22.44}$$

The symmetry conditions (22.43) here read

$$\mathscr{L}_{\mathbf{X}}\boldsymbol{\alpha}_1 = \lambda_1\boldsymbol{\alpha}_1 + \lambda_2\boldsymbol{\alpha}_2 \tag{22.45}$$

and

$$\mathscr{L}_{\mathbf{X}}\boldsymbol{\alpha}_2 = \lambda_3\boldsymbol{\alpha}_1 + \lambda_4\boldsymbol{\alpha}_2, \tag{22.46}$$

where the λ_i are functions of x, y, and p.

To evaluate these symmetry conditions, that is, to determine the components ξ, η, and η' of the symmetry generator

$$\mathbf{X} = \xi\frac{\partial}{\partial x} + \eta\frac{\partial}{\partial y} + \eta'\frac{\partial}{\partial p}, \tag{22.47}$$

it is best to use the definition (22.25) of the Lie derivative. If we substitute

$$\begin{aligned} \mathbf{X} \lrcorner \boldsymbol{\alpha}_1 &= \left(\xi\frac{\partial}{\partial x} + \eta\frac{\partial}{\partial y} + \eta'\frac{\partial}{\partial p}\right) \lrcorner (\mathrm{d}y - p\,\mathrm{d}x) = -p\xi + \eta, \\ \mathbf{X} \lrcorner \mathrm{d}\boldsymbol{\alpha}_1 &= \left(\xi\frac{\partial}{\partial x} + \eta\frac{\partial}{\partial y} + \eta'\frac{\partial}{\partial p}\right) \lrcorner (-\mathrm{d}p \wedge \mathrm{d}x) = \xi\mathrm{d}p - \eta'\,\mathrm{d}x \end{aligned} \tag{22.48}$$

into $\mathcal{L}_{\mathbf{X}}\boldsymbol{\alpha}_1 = \mathbf{X} \lrcorner \mathrm{d}\boldsymbol{\alpha}_1 + \mathrm{d}(\mathbf{X} \lrcorner \boldsymbol{\alpha}_1)$, we obtain from (22.45) the condition

$$-\eta' \,\mathrm{d}x + \mathrm{d}\eta - p\,\mathrm{d}\xi = \lambda_1(\mathrm{d}y - p\,\mathrm{d}x) + \lambda_2(\mathrm{d}p - \omega\,\mathrm{d}x) \tag{22.49}$$

or

$$(-\eta' + \eta_{,x} - p\xi_{,x})\,\mathrm{d}x + (\eta_{,y} - p\xi_{,y})\,\mathrm{d}y + (\eta_{,p} - p\xi_{,p})\,\mathrm{d}p$$
$$= (-\lambda_1 p - \lambda_2\omega)\,\mathrm{d}x + \lambda_1\,\mathrm{d}y + \lambda_2\,\mathrm{d}p. \tag{22.50}$$

Comparing the coefficients of $\mathrm{d}y$ and $\mathrm{d}p$ on the two sides of this equation, λ_1 and λ_2 can easily be read off, and using these results, we finally obtain

$$\eta' = \eta_{,x} + p\eta_{,y} + \omega\eta_{,p} - p(\xi_{,x} + p\xi_{,y} + \omega\xi_{,p}). \tag{22.51}$$

If we replace p by y' and compare this expression for η' with that given in equation (12.9) or (12.16), we see that the two coincide: the condition (22.51) imposed on η' is exactly the extension law of a dynamical symmetry; compare Chapter 12.

In a similar way we can evaluate the second symmetry condition (22.46). Since $\boldsymbol{\alpha}_2$ and $\mathcal{L}_{\mathbf{X}}\boldsymbol{\alpha}_2$ are 1-forms, λ_3 and λ_4 are functions (as λ_1 and λ_2 were) that turn out to be

$$\lambda_3 = \eta'_{,y} - \omega\xi_{,y}, \qquad \lambda_4 = \eta'_{,p} - \omega\xi_{,p}, \tag{22.52}$$

and (22.46) then yields

$$\xi\omega_{,x} + \eta\omega_{,y} + \eta'\omega_{,p} - \eta'_{,x} - p\eta'_{,y} - \omega\eta'_{,p}$$
$$+ \omega(\xi_{,x} + \eta\xi_{,y} + \omega\xi_{,p}) = 0. \tag{22.53}$$

This is exactly the symmetry condition (12.17) for a dynamical symmetry if $y' = p$ and the definitions (12.14) and (12.16) of $\mathbf{A}$ and η' are taken into account.

So we see that for ordinary differential equations the definition (22.43) of a symmetry leads to the dynamical symmetries. Lie point symmetries and contact symmetries are contained as subcases that occur when the necessary constraints are imposed on the components (ξ, η) of the vector $\mathbf{X} = \xi\,\partial/\partial x + \eta\,\partial/\partial y + \cdots$; compare Sections 11.2 and 12.2.

A (quasi-) *linear partial differential equation* such as the heat conduction equation

$$\boldsymbol{\alpha}_1 = \mathrm{d}u - p\,\mathrm{d}x - q\,\mathrm{d}t, \qquad \boldsymbol{\alpha}_2 = \mathrm{d}p \wedge \mathrm{d}t - q\,\mathrm{d}x \wedge \mathrm{d}t, \tag{22.54}$$
$$\boldsymbol{\alpha}_3 = -\,\mathrm{d}p \wedge \mathrm{d}x - \mathrm{d}q \wedge \mathrm{d}t$$

typically contains only *one* 1-form in its representation by forms, which is

$\boldsymbol{\alpha}_1$. For a general quasi-linear equation with more than two independent variables this form would read

$$\boldsymbol{\alpha}_1 = \mathrm{d}u - p_i\,\mathrm{d}x^i. \tag{22.55}$$

Since the Lie derivative of a 1-form is again a 1-form, only $\boldsymbol{\alpha}_1$ can appear on the right side of the symmetry condition for $\boldsymbol{\alpha}_1$, that is, we have

$$\mathscr{L}_{\mathbf{X}}\boldsymbol{\alpha}_1 = \lambda\boldsymbol{\alpha}_1. \tag{22.56}$$

With

$$\mathbf{X} = \xi^i\frac{\partial}{\partial x^i} + \eta\frac{\partial}{\partial u} + \eta_i\frac{\partial}{\partial p_i} \tag{22.57}$$

and therefore

$$\mathbf{X} \lrcorner\, \boldsymbol{\alpha}_1 = \eta - \xi^i p_i, \qquad \mathbf{X} \lrcorner\, \mathrm{d}\boldsymbol{\alpha}_1 = \xi^i\,\mathrm{d}p_i - \eta_i\,\mathrm{d}x^i, \tag{22.58}$$

the symmetry condition (22.56) reads in full

$$\mathrm{d}\eta - p_i\,\mathrm{d}\xi^i - \eta_i\,\mathrm{d}x^i = \lambda(\mathrm{d}u - p_i\,\mathrm{d}x^i). \tag{22.59}$$

Since the independent variables used here are u, x^i, and p_i, (22.59) shows that $\mathrm{d}\eta - p_i\,\mathrm{d}\xi^i$ does not contain terms in $\mathrm{d}p_a$. From the coefficients of $\mathrm{d}u$, λ can be determined to be $\lambda = \eta_{,u} - p_i\xi^i{}_{,u}$, and with this (22.59) yields

$$\eta_i = \eta_{,i} + p_i\eta_{,u} - p_a(\xi^a{}_{,i} - p\xi^a{}_{,u}), \tag{22.60}$$

which because of $p_i = u_{,i}$ is equivalent to the extension law (16.40) for Lie point symmetries and the analogous law (21.8) for contact symmetries [note that (22.59) implies that (21.9) is also satisfied]: for *one* (quasi-linear) partial differential equation, the symmetries defined by (22.30) are exactly the contact and point symmetries.

To evaluate the rest of the symmetry conditions, that is, those arising from the Lie derivatives of $\boldsymbol{\alpha}_2$ and $\boldsymbol{\alpha}_3$ for the heat conduction equation (22.54), is a more difficult task. It turns out, however, that no genuine contact symmetries exist and that the Lie point symmetries are exactly those given in (16.64).

We conclude this section with a few remarks on similarity solutions and conditional symmetries; compare Chapter 18.

Looking for similarity solutions means adding to the differential equations

$$\boldsymbol{\alpha}_i = 0 \tag{22.61}$$

a further restriction which in essence says that the solution does not depend on the variable corresponding to the symmetry generator **X**. This is achieved by imposing the conditions

$$\boldsymbol{\sigma}_i \equiv \mathbf{X} \lrcorner \boldsymbol{\alpha}_i = 0. \tag{22.62}$$

If, for example, applied to $\boldsymbol{\alpha}_1$ given by (22.55), equation (22.62) yields

$$\boldsymbol{\sigma}_1 = \eta - \xi^i p_i = 0,$$

which agrees with (18.7) in the case of a single dependent variable.

Looking for conditional symmetries means looking for vectors **X** that are not symmetries of the set $\{\boldsymbol{\alpha}_i\}$, but only of the enlarged set $\{\boldsymbol{\omega}_i\} = \{\boldsymbol{\alpha}_i, \boldsymbol{\sigma}_i\}$, that is, vectors **X** that satisfy

$$\mathscr{L}_{\mathbf{X}} \boldsymbol{\omega}_i = \boldsymbol{\lambda}_i^k \wedge \boldsymbol{\omega}_k, \tag{22.63}$$

the $\boldsymbol{\sigma}_i$ occurring also on the right side of (22.63).

22.5 Exercises

1 Show that in an n-dimensional space no nonzero p-forms with $p > n$ exist.

2 Show that the commutation rule (22.12) holds.

3 Give the system (16.30), that is, $u_{,y} - v_{,x} = 0$, $uu_{,x} + v_{,y} = 0$, in terms of a closed set of forms.

4 Which partial differential equation is equivalent to the set

$$\boldsymbol{\alpha}_1 = \mathrm{d}t \wedge \mathrm{d}v \wedge \mathrm{d}w - \mathrm{d}y \wedge \mathrm{d}u \wedge \mathrm{d}w - z\,\mathrm{d}u \wedge \mathrm{d}v \wedge \mathrm{d}w,$$

$$\boldsymbol{\alpha}_2 = \mathrm{d}t \wedge \mathrm{d}u + \mathrm{d}y \wedge \mathrm{d}v + \mathrm{d}z \wedge \mathrm{d}w,$$

$$\boldsymbol{\alpha}_3 = \mathrm{d}x - t\,\mathrm{d}u - y\,\mathrm{d}v - z\,\mathrm{d}w$$

of forms?

5 Show that the set $\{\boldsymbol{\alpha}_i, \boldsymbol{\sigma}_i \equiv X \lrcorner \boldsymbol{\alpha}_i\}$ is closed if $\boldsymbol{\alpha}_i$ is closed and if **X** is a symmetry generator of $\boldsymbol{\alpha}_i$.

23

Lie–Bäcklund transformations

23.1 Why study more general transformations and symmetries?

For partial differential equations, we have so far studied Lie point transformations and contact transformations and the invariance of the differential equations with respect to groups of such transformations. Lie point transformations are (invertible) transformations in the space (x^n, u^α) of independent and dependent variables, which can then be extended to the derivatives, and the infinitesimal generator of a group of point transformations is the extension of

$$\mathbf{X} = \xi^i(x^n, u^\beta)\frac{\partial}{\partial x^i} + \eta^\alpha(x^n, u^\beta)\frac{\partial}{\partial u^\alpha}, \tag{23.1}$$

where ξ^i and η^α are functions of the indicated variables. Contact transformations – which exist only in the case of a single dependent variable u – are transformations in the space $(x^n, u, u_{,n})$ of independent and dependent variables and their first derivatives, which can then be extended to higher derivatives, and the infinitesimal generator of a group of contact transformations is the extension of

$$\begin{gathered}\mathbf{X} = \xi^i(x^n, u, u_{,n})\frac{\partial}{\partial x^i} + \eta(x^n, u, u_n)\frac{\partial}{\partial u} + \eta_i(x^n, u, u_{,n})\frac{\partial}{\partial u_{,i}},\\ \xi^i = \frac{\partial \Omega}{\partial u_{,i}}, \quad \eta = u_{,n}\frac{\partial \Omega}{\partial u_{,n}} - \Omega, \quad \eta_i = -\left(u_{,i}\frac{\partial \Omega}{\partial x} + \frac{\partial \Omega}{\partial x^i}\right),\end{gathered} \tag{23.2}$$

where $\Omega(x^i, u, u_{,i})$ is an arbitrary function. Note that η_i need not necessarily be given since it can be calculated from ξ^i and η via the usual extension law

$$\eta_n = \frac{\mathrm{D}}{\mathrm{D}x^n}\eta - u_{,i}\frac{\mathrm{D}\xi^i}{\mathrm{D}x^n}. \tag{23.3}$$

If we now want to generalize these types of transformations, it is obvious how this could be done: we have to admit transformations that mix independent and dependent variables *and* their derivatives up to at least second order. Or in the language of the generators **X**: we should admit generators that are extensions of

$$\mathbf{X} = \xi^i(x^n, u^\beta, u^\beta{}_{,n} u^\beta{}_{,nm}, \dots)\frac{\partial}{\partial x^i} + \eta^\alpha(x^n, u^\beta, u^\beta{}_{,n}, u^\beta{}_{,nm}, \dots)\frac{\partial}{\partial u^\alpha}, \tag{23.4}$$

where ξ^i and η^α depend on the higher derivatives or where – if only first derivatives appear in ξ^i and η^α – the terms with second derivatives naturally arising in the first extension η^α_i do not cancel out as they do for contact transformations. Before discussing how a generalization can be made without running into contradictions, we collect a few arguments that indicate that such a generalization is natural and makes sense.

The first argument is the analogy with the case of ordinary differential equations, where point and contact transformations and symmetries were successfully generalized to dynamical symmetries.

The second argument in favour of a generalization arises from a quite common practice in dealing with differential equations. It may happen that a system

$$H_A(x^i, u^\alpha{}_{,i}, u^\alpha{}_{,ik}, \dots) = 0 \tag{23.5}$$

admits the introduction of a potential v in the sense that functions f_n exist such that

$$v_{,n} = f_n(x^i, v, u^\alpha, u^\alpha{}_{,i}, \dots) \tag{23.6}$$

holds, and that the integrability conditions $v_{,nm} = v_{,mn}$ of (23.6) are satisfied identically because of (23.5) and (23.6). In terms of this new variable v and the (perhaps also modified) remaining variables $\hat{u}^\alpha$, the system (23.5) will then read

$$G_A(x^i, v, \hat{u}^\alpha, v_{,i}, v_{,ik}, \hat{u}^\alpha{}_{,i}, \hat{u}^\alpha{}_{,ik}, \dots) = 0, \tag{23.7}$$

and if the transformation (23.6) was successful, it will be easier to treat than the original system (23.5). Suppose now that the system $G_A = 0$ admits a Lie point symmetry in $(x^i, v, \hat{u}^\alpha)$-space. Due to the nonlocal character of the transformation (23.6) between v and the u^α, this symmetry transformation will in general involve derivatives of the u^α when expressed in terms of x^i and u^α.

To illustrate this by an example, we take the Burgers equation

$$u_{,t} + 2uu_{,x} - u_{,xx} = 0. \tag{23.8}$$

It admits the introduction of a new dependent variable v (the potential) by means of the two equations

$$\begin{aligned} v_{,x} &= -uv, \\ v_{,t} &= -vu_{,x} + vu^2, \end{aligned} \tag{23.9}$$

which characterize the Cole–Hopf transformation. The integrability condition $v_{,xt} = v_{,tx}$ gives

$$v(u_{,t} + 2uu_{,x} - u_{,xx}) = v_{,x}(u_{,x} - u^2) - uv_{,t}, \tag{23.10}$$

which is satisfied as the left side vanishes because of (23.8) and the right side because of (23.9). An immediate consequence of the system (23.9) is that the potential v satisfies the heat conduction equation

$$v_{,t} - v_{,xx} = 0. \tag{23.11}$$

If, on the other hand, we use the first of the equations (23.9), namely,

$$u = -v_{,x}/v, \tag{23.12}$$

to formulate the Burgers equation in terms of v, we see that it is identically satisfied provided v satisfies the heat conduction equation (23.11). So the introduction of the potential v by (23.9) transforms the nonlinear Burgers equation into the linear heat conduction equation.

Among the many symmetries (16.64) of the heat conduction equation (23.11) is that with generator

$$\mathbf{X} = g(x,t)\frac{\partial}{\partial v} + g_{,i}\frac{\partial}{\partial v_{,i}} + \cdots, \qquad g_{,t} - g_{,xx} = 0, \tag{23.13}$$

which is essentially due to the linearity of that equation. Because of the equivalence of the heat and Burgers equations, it should show up also as a symmetry of the Burgers equation. If we simply transform the generator (23.13) to the variables (x, t, u) by using (23.12), we initially obtain

$$\mathbf{X} = (\mathbf{X}u)\frac{\partial}{\partial u} \tag{23.14}$$

with

$$\mathbf{X}u = \left(g\frac{\partial}{\partial v} + g_{,i}\frac{\partial}{\partial v_{,i}}\right)\left(-\frac{v_{,x}}{v}\right) = -\frac{1}{v}(g_{,x} + ug). \tag{23.15}$$

To give $\mathbf{X}u$ entirely in terms of u (and its derivatives), we have to know v in terms of u. Because of the defining equations (23.9), v can be given by a line integral as

$$\ln v(x,t) = -\int_{x_0,t_0}^{x,t_0} u(x',t_0)\,\mathrm{d}x' + \int_{x,t_0}^{x,t} [u^2(x,t') - u_{,x}(x,t')]\,\mathrm{d}t', \tag{23.16}$$

so that $v(x,t)$ depends not only on u at the same point (x,t), but on the values of u along the whole line of integration, that is, on u and all its derivatives. So if we want to have (23.13) among the symmetries of the Burgers equation, we have to admit a rather wide class of transformations.

The third argument supporting a more general concept of symmetries runs as follows. In Section 12.3 we observed that the most general symmetries of ordinary differential equations, the dynamical symmetries, can be interpreted as arising from simple changes of the initial values of the solution. This idea can be transferred to partial differential equations. Imagine that the solution of a (set of) partial differential equation(s) is given in terms of its (arbitrary) initial, or boundary, values. Any change in these initial values will produce another solution; it maps solutions into solutions and is, therefore, a symmetry transformation. Due to the nonlocal relation between initial values and solutions, the explicit form of such a symmetry transformation can be very complicated even for rather simple changes in the initial data. Of course, this last argument not only supports a rather general form of symmetries, but it also indicates that *every* partial differential equation should have *many* generalized symmetries, rather as every ordinary differential equation always admits dynamical symmetries.

23.2 Finite order generalizations do not exist

Before we come to the proper generalization of Lie point (or contact) transformations and symmetries, we want to discuss a possibility that will finally be ruled out. This possibility is to admit transformations that are essentially point transformations in the space (x^i, u^α, $u^\alpha{}_{,i_1}$, $u^\alpha{}_{,i_1 i_2}, \dots, u_{,i_1 \cdots i_k}$) of independent and dependent variables and their derivatives *up to a fixed order k*. For the infinitesimal generators

$$\mathbf{X} = \xi^n(x^i, u^\beta, u^\beta{}_{,i_1}, \dots, u^\beta{}_{,i_1 \cdots i_k}) \frac{\partial}{\partial x^n}$$

$$+ \eta^\alpha(x^i, u^\beta, u^\beta{}_{,i_1}, \dots, u^\beta{}_{,i_1 \cdots i_k}) \frac{\partial}{\partial u^\alpha} \tag{23.17}$$

of these transformations, the above restriction means that in none of the coefficients $\eta^\alpha_{i_1}, \dots, \eta^\alpha_{i_1 \cdots i_k}$ of the extension of that generator up to the kth order are derivatives with order $m > k$ permitted. This condition on the extension of $\mathbf{X}$ is a rather stringent one. We already showed in Section 21.1 that for $k = 1$ it can be fulfilled only by extended point transformations (if there are several dependent variables u^α) or by contact transformations (if there is only one dependent variable u). The proof given there can be generalized to the case $k > 1$. If we assume that $\eta^\alpha_{i_1 \cdots i_k}$ does not depend on $u^\beta{}_{,j_1 \cdots j_{kn}}$, then because of the extension law

$$\eta^\alpha_{i_1 \cdots i_k} = \frac{\mathrm{D}\eta^\alpha_{i_1 \cdots i_{k-1}}}{\mathrm{D}x_{i_k}} - u^\alpha{}_{,i_1 \cdots k_{k-1} r} \frac{\mathrm{D}\xi^r}{\mathrm{D}x_{i_k}}, \tag{23.18}$$

the coefficients $\eta^\alpha_{i_1 \cdots i_{k-1}}$ and ξ^r cannot depend on derivatives of order k but at most on derivatives of order $k - 1$. Repeating this argument for the coefficients $\eta^\alpha_{i_1 \cdots i_{k-1}}$, one finally ends up with the statement that the coefficients η^α and ξ^r do not depend on the first derivatives; that is, the transformation supposed to be of order k is only an extended point transformation (or contact transformation in the case of one dependent variable).

This means that genuine finite order transformations do not exist. If the coefficients of the kth extension $\mathbf{X}^{(k)}$ ($k > 0$) of a generator depend only on derivatives of order $m \leqslant k$, then the transformation is a point or a contact transformation (the same is true for transformations that are not members of a one-parameter group and cannot be treated by the discussion of generators).

In the case of ordinary differential equations there was an escape from this law; when looking only for symmetries (and transformations) of a given differential equation, we could use that differential equation to eliminate the unwelcome higher derivatives. The resulting transformations then were the dynamical symmetries treated in Chapters 11–13.

Is there a similar escape in the case of partial differential equations? To get an impression, we consider the simple case of one second order differential equation

$$H(x^n, u, u_{,n}, u_{,nm}) = 0 \tag{23.19}$$

for one dependent variable u. If we look for symmetries

$$\mathbf{X} = \xi^n(x^a, u, u_{,a})\frac{\partial}{\partial x^n} + \eta(x^a, u, u_{,a})\frac{\partial}{\partial u} + \eta_n\frac{\partial}{\partial u_{,n}}, \tag{23.20}$$

where the η_n do not depend on the second derivatives, then because of the extension law (23.13), we have to guarantee that the first term in

$$\eta_n = u_{,nm}\left(\frac{\partial\eta}{\partial u_{,m}} - u_a\frac{\partial\xi^a}{\partial u_{,m}}\right) + \text{terms in } u_{,m}, u, x^m \tag{23.21}$$

vanishes *identically* due to the differential equation (23.19). These are N equations, where N is the number of the independent variables x^n, and for $N > 1$ it is impossible to satisfy all of them by using $H = 0$ even if several of the N brackets appearing in (23.21) vanish (if all of them vanish, we have a contact symmetry). This procedure can only work for $N = 1$, that is, for ordinary differential equations. For several dependent variables and (therefore) several differential equations or for higher order differential equations, the answer will be the same: finite order generalizations of the dynamical symmetries for partial differential equations do not exist.

23.3 Lie–Bäcklund transformations and their infinitesimal generators

The discussion of the two preceding sections shows that in order to genuinely generalize point and contact transformations we must admit derivatives of arbitrarily high order on the right side of the transformation laws

$$\begin{aligned} \tilde{x}^n &= \tilde{x}^n(x^i, u^\beta, u^\beta{}_{,i}, u^\beta{}_{,ik}, \ldots), \\ \tilde{u}^\alpha &= \tilde{u}^\alpha(x^i, u^\beta, u^\beta{}_{,i}, u^\beta{}_{,ik}, \ldots), \\ \tilde{u}^\alpha{}_{,n} &= \tilde{u}^\alpha{}_{,n}(x^i, u^\beta, u^\beta{}_{,i}, u^\beta{}_{,ik}, \ldots), \\ &\vdots \end{aligned} \tag{23.22}$$

To admit the interpretation as derivatives, the quantities $\tilde{u}^\alpha{}_{,n}$ must of course obey $\tilde{u}^\alpha{}_{,n} = \partial\tilde{u}^\alpha/\partial\tilde{x}^n$, and analogous conditions must be true for the transformation of the higher derivatives. This is often formulated by saying that the transformations (23.22) must leave invariant the contact (or tangency) conditions

$$
\begin{aligned}
&du^\alpha - u^\alpha{}_{,i}\,dx^i = 0,\\
&du^\alpha{}_{,n} - u^\alpha{}_{,ni}\,dx^i = 0,\\
&\qquad\vdots
\end{aligned}
\tag{23.23}
$$

Transformations (23.22), which satisfy these conditions, are called generalized transformations or Lie–Bäcklund transformations. We shall prefer the latter name (despite the danger of a mix-up with the different concept of a Bäcklund transformation: compare Section 23.5) since we feel that "generalized" has too diffuse a meaning.

Lie–Bäcklund transformations generalize Lie point and contact transformations and contain them as special subcases. For a genuine generalization, not all of the transformations (23.22) need contain derivatives of arbitrarily high order; it suffices that the kth derivatives $\tilde{u}^\alpha{}_{,n_1\cdots n_k}$ depend on the derivatives of order at least $k+1$. In the more interesting and more difficult cases, derivatives of arbitrarily high order will occur in each of the transformations (23.22). In these cases it will often prove more appropriate or even necessary to introduce potentials and use integrals (as well as partial derivatives) as the variables on which Lie–Bäcklund transformations act.

In general, a Lie–Bäcklund transformation will not depend on an arbitrary parameter and will not be a member of a (one-parameter) group. But if such an one-parameter group exists, then all the transformations (23.22) will depend on the group parameter ε,

$$
\begin{aligned}
\tilde{x}^n &= \tilde{x}^n(x^i, u^\beta, u^\beta{}_{,i}, \ldots; \varepsilon) = x^n + \varepsilon\xi^n(x^i, u^\beta, u^\beta{}_{,i}, \ldots) + \cdots,\\
\tilde{u}^\alpha &= \tilde{u}^\alpha(x^i, u^\beta, u^\beta{}_{,i}, \ldots; \varepsilon) = u^\alpha + \varepsilon\eta^\alpha(x^i, u^\beta, u^\beta{}_{,i}, \ldots) + \cdots,
\end{aligned}
\tag{23.24}
$$

and so on. We can define the infinitesimal generator $\mathbf{X}$ of these transformations by

$$
\mathbf{X} = \xi^n \frac{\partial}{\partial x^n} + \eta^\alpha \frac{\partial}{\partial u^\alpha} + \eta^\alpha_i \frac{\partial}{\partial u^\alpha{}_{,i}} + \eta^\alpha_{ik} \frac{\partial}{\partial u^\alpha{}_{,ik}} + \cdots,
\tag{23.25}
$$

where

$$
\begin{aligned}
\xi^n(x^i, u^\beta, u^\beta{}_{,i}, u^\beta{}_{,ik}, \ldots) &= \left.\frac{\partial \tilde{x}^n}{\partial \varepsilon}\right|_{\varepsilon=0},\\
\eta^\alpha(x^i, u^\beta, u^\beta{}_{,i}, u^\beta{}_{,ik}, \ldots) &= \left.\frac{\partial \tilde{u}^\alpha}{\partial \varepsilon}\right|_{\varepsilon=0},
\end{aligned}
\tag{23.26}
$$

and so on, holds and the extension to the derivatives is given by, for example,

$$\eta^\alpha_i = \frac{\mathrm{D}}{\mathrm{D}x^i}\eta^\alpha - u^\alpha{}_{,n}\frac{\mathrm{D}\xi^n}{\mathrm{D}x^i}, \tag{23.27}$$

with

$$\frac{\mathrm{D}}{\mathrm{D}x^i} = \frac{\partial}{\partial x^i} + u^\beta{}_{,i}\frac{\partial}{\partial u^\beta} + u^\beta{}_{,ki}\frac{\partial}{\partial u^\beta{}_{,k}} + \cdots; \tag{23.28}$$

compare Section 15.1, where all this was derived in detail for the case of point transformations. As in that case, everything is known if we have to hand the functions $\tilde{x}^n$, $\tilde{u}^\alpha$ or ξ^n, η^α; the extension to the derivatives is a straightforward program.

If the generator **X** is given, the finite transformations can in principle be obtained by integrating the infinite system

$$\frac{\mathrm{d}\tilde{x}^n}{\mathrm{d}\varepsilon} = \xi^n(\tilde{x}^i, \tilde{u}^\alpha, \tilde{u}^\alpha{}_{,i}, \ldots), \quad \frac{\mathrm{d}\tilde{u}^\alpha}{\mathrm{d}\varepsilon} = \eta^\alpha, \quad \frac{\mathrm{d}\tilde{u}^\alpha{}_{,n}}{\mathrm{d}\varepsilon} = \eta^\alpha_n, \quad \text{etc.}, \tag{23.29}$$

of ordinary differential equations with initial values $\tilde{x}^n = x^n$, $\tilde{u}^\alpha = u^\alpha, \ldots,$ for $\varepsilon = 0$. By taking the trajectories of these finite transformations as coordinate lines, that is, by performing a suitable Lie–Bäcklund transformation, any generator **X** can in principle be transformed into its normal form

$$\mathbf{X} = \frac{\partial}{\partial \psi} \tag{23.30}$$

(which is the generator of a translation along the trajectories).

So formally there seem to be no difficulties in generalizing all these concepts from point transformations to Lie–Bäcklund transformations. The difficulties arise when we have to ensure that all the processes involved are well defined and can be performed rigorously. We cannot and will not do this here but rather proceed in an intuitive fashion as we often did earlier. Even then the technical problems in dealing with Lie–Bäcklund transformations and symmetries will be great enough.

23.4 Examples of Lie–Bäcklund transformations

1. The one-parameter group of transformations

$$\tilde{x} = x, \qquad \tilde{y} = \sum_{n=0}^{\infty} \frac{\varepsilon^n}{n!} y^{(n)} \tag{23.31}$$

are Lie–Bäcklund transformations. Because of, for example, $\mathrm{d}\tilde{y}/\mathrm{d}\varepsilon|_{\varepsilon=0} = y'$, the associated generator is

$$\mathbf{X} = y'\frac{\partial}{\partial y} + y''\frac{\partial}{\partial y'} + y'''\frac{\partial}{\partial y''} + \cdots. \tag{23.32}$$

If applied to a function $y(x)$, the transformation (23.31) yields

$$\tilde{y}(x) = \sum_{n=0}^{\infty} \frac{\varepsilon^n}{n!} y^{(n)}(x), \qquad \tilde{x} = x, \tag{23.33}$$

and obviously this can be given the closed form

$$\tilde{y}(x) = y(x+\varepsilon), \qquad \tilde{x} = x. \tag{23.34}$$

This form shows that the net result of the above Lie–Bäcklund transformation is the same as that of a simple translation in x! We shall get a better understanding of this correspondence in Section 24.2.

2. To determine the group of Lie–Bäcklund transformations associated with a given generator of the special form

$$\mathbf{X} = \eta^\alpha \frac{\partial}{\partial u^\alpha} + \eta_i^\alpha \frac{\partial}{\partial u_i^\alpha} + \cdots \tag{23.35}$$

(all the ξ^n are zero), it suffices to determine the functions $\tilde{u}^\alpha$ from

$$\frac{\mathrm{d}\tilde{u}^\alpha}{\mathrm{d}\varepsilon} = \eta^\alpha(x^i, \tilde{u}^\beta, \tilde{u}^\beta{}_{,i}, \ldots), \qquad \tilde{u}^\alpha|_{\varepsilon=0} = u^\alpha, \tag{23.36}$$

as we obviously have $\tilde{x}^i = x^i$, and the rest of the transformed quantities $\tilde{u}^\alpha{}_{,i}, \ldots$ can be computed by means of the extension rules. To solve $\mathrm{d}\tilde{u}^\alpha/\mathrm{d}\varepsilon = \eta^\alpha$, we can think of the $\tilde{u}^\alpha$ as functions of the x^i and of ε and regard these equations as partial differential equations with prescribed initial values,

$$\frac{\partial \tilde{u}^\alpha}{\partial \varepsilon} = \eta^\alpha(x^i, \tilde{u}^\beta, \tilde{u}^\beta{}_{,i}, \ldots), \qquad \tilde{u}^\alpha|_{\varepsilon=0} = u^\alpha(x^i). \tag{23.37}$$

That is, we have to solve a Cauchy problem for an autonomous system (ε does not appear explicitly on the right side). Conversely, any such system can be thought of as defining a Lie–Bäcklund transformation.

If we apply this idea to the generator

$$\mathbf{X} = \eta\frac{\partial}{\partial u} + \cdots = (u_{,xx} + u_{,yy})\frac{\partial}{\partial u} + \cdots, \tag{23.38}$$

we are led to the initial-value problem for the two-dimensional heat conduction equation,

$$\frac{\partial \tilde{u}}{\partial \varepsilon} = \frac{\partial^2 \tilde{u}}{\partial x^2} + \frac{\partial^2 \tilde{u}}{\partial y^2}, \qquad \tilde{u}(x, y, \varepsilon)|_{\varepsilon=0} = u(x, y). \tag{23.39}$$

Its solution for $\varepsilon > 0$ is

$$\tilde{u}(x, y, \varepsilon) = \frac{1}{4\pi\varepsilon}\int \exp\left\{-\frac{1}{4\varepsilon}[(x-a)^2 + (y-b)^2]\right\} u(a, b)\,\mathrm{d}a\,\mathrm{d}b. \tag{23.40}$$

Together with $\tilde{x} = x$, $\tilde{y} = y$, this is the finite Lie–Bäcklund transformation corresponding to the generator (23.38). To see that derivatives really appear on the right side of (23.40), one should imagine that $u(a, b)$ is replaced by its Taylor series at the point (x, y).

3. An example of a Lie–Bäcklund transformation that is *not* a member of a one-parameter group is

$$\tilde{x} = x, \qquad \tilde{y} = y, \qquad \tilde{u} = -u_{,x}/u, \tag{23.41}$$

which coincides with the first line of the Cole–Hopf transformation (23.9). To make the transformation (23.41) have a unique inverse, an additional constraint has to be imposed on u.

23.5 Lie–Bäcklund versus Bäcklund transformations

Lie–Bäcklund transformations are transformations in the space of independent and dependent variables and their derivatives. They may be applied to differential equations, but a given Lie–Bäcklund transformation does not depend on, or in its definition refer to, a particular differential equation.

Bäcklund transformations, which are widely used nowadays in soliton physics, have a rather different status. They too are transformations in which the (independent and dependent) variables as well as their derivatives are involved, but each of them makes sense, and is well defined, only for an associated special set of partial differential equations and its image under this transformation (if the image coincides with the original set, the transformation is sometimes called an auto-Bäcklund transformation). In the case of two independent variables (x, y) and one dependent variable u, a Lie–Bäcklund transformation is defined by prescribing *three* functions, namely,

$$\tilde{x} = \tilde{x}(x, y, u, u_{,x}, u_{,y}, \ldots), \qquad \tilde{y} = \tilde{y}(\cdots), \qquad \tilde{u} = \tilde{u}(\cdots), \tag{23.42}$$

from which the transformations of the derivatives, for example,

$$\tilde{u}_{,x} = \tilde{u}_{,x}(x, y, u, u_{,x}, u_{,y}, \ldots), \qquad \tilde{u}_{,y} = \tilde{u}_{,y}(\cdots), \tag{23.43}$$

can in principle be determined by extension. A Bäcklund transformation typically prescribes *four* relations,

$$F_A(x, y, u, u_{,x}, u_{,y}; \tilde{x}, \tilde{y}, \tilde{u}, \tilde{u}_{,x}, \tilde{u}_{,y}) = 0, \qquad A = 1, \ldots, 4, \tag{23.44}$$

between the old and the new variables and their first derivatives. Since in general there is no additional degree of freedom to account for this one surplus relation, the four relations (23.44) can hold only if restricted to special sets of $(x, y, u, u_{,x}, u_{,y})$ and their images, that is, if restricted to solutions of a special differential equation that can be obtained by studying the four equations (23.44). So Bäcklund transformations are essentially transformations in the space of solutions of differential equations.

An example of a Bäcklund transformation is the transformation (23.9), the Cole–Hopf transformation, which on including the transformations for x and y in full reads

$$\begin{aligned} &\tilde{x} = x, \qquad \tilde{y} = y, \qquad \tilde{u} = -u_{,x}/u, \\ &\tilde{u}_{,x} = (u_{,x}/u)^2 - u_{,t}/u. \end{aligned} \tag{23.45}$$

It is well defined only if u is a solution of the heat conduction equation and $\tilde{u}$ a solution to the Burgers equation; compare Section 23.1. When comparing the Lie–Bäcklund transformation (23.41) with the Bäcklund transformation (23.45), one clearly sees how the introduction of a fourth independent transformation law decisively changes the situation.

It is not always possible to give a Bäcklund transformation an explicit form, as in (23.45), where the new variables are given in terms of the old ones, as the example of the most famous Bäcklund transformation,

$$\begin{aligned} &\tfrac{1}{2}(\tilde{u}_{,x} - u_{,x}) = a \sin\left(\frac{\tilde{u} + u}{2}\right), \\ &\frac{a}{2}(\tilde{u}_{,y} + u_{,y}) = \sin\left(\frac{\tilde{u} - u}{2}\right), \\ &\tilde{x} = x, \qquad \tilde{y} = y, \qquad a = \text{const.}, \end{aligned} \tag{23.46}$$

shows. By differentiating the first line with respect to y and the second line with respect to x, one obtains

$$\tilde{u}_{,xy} - u_{,xy} = \sin\tilde{u} - \sin u, \qquad \tilde{u}_{,xy} + u_{,xy} = \sin\tilde{u} + \sin u; \qquad (23.47)$$

the functions $\tilde{u}(x, y)$ and $u(x, y)$ have to obey the same differential equation

$$u_{,xy} = \sin u. \qquad (23.48)$$

The transformation (23.46) is thus an auto-Bäcklund transformation of the sine–Gordon equation (23.48) [a different form of this equation has been given in (16.65)]. The appearance of an arbitrary parameter a in the Bäcklund transformation is essential to its utility in solving the sine–Gordon equation.

23.6 Exercises

1 Determine the finite transformation corresponding to the generator (23.32) by using the method explained in Example 2 of Section 23.4.

2 Do the Bäcklund transformations (23.46) form a one-parameter group?

24

Lie–Bäcklund symmetries and how to find them

24.1 The basic definitions

If we apply a Lie–Bäcklund transformation to a set

$$H_A(x^i, u^\alpha, u^\alpha{}_{,i}, \ldots) = 0 \tag{24.1}$$

of differential equations, the result will in general be a different set $G_A = 0$ of differential equations, which moreover will be of different, possibly even infinite, order. When we want to define a Lie–Bäcklund symmetry as an invariance property of a system of differential equations, we have to take into account this possible change of order. This is done by considering not only the set $H_A = 0$ but also all its differential consequences

$$\begin{aligned} D_i H_A &= 0, \\ D_i D_j H_A &= 0, \\ &\vdots \end{aligned} \tag{24.2}$$

where

$$D_i = \frac{\mathrm{D}}{\mathrm{D}x^i} = \frac{\partial}{\partial x^i} + u^\alpha{}_{,i} \frac{\partial}{\partial u^\alpha} + u^\alpha{}_{,ik} \frac{\partial}{\partial u^\alpha{}_{,k}} + \cdots . \tag{24.3}$$

We shall then say that a Lie–Bäcklund transformation is a Lie–Bäcklund symmetry if it leaves invariant the system $H_A = 0$ together with all its differential consequences (24.2).

For practical purposes, this is a rather unwieldy definition, and we immediately restrict ourselves to Lie–Bäcklund transformations that are members of a one-parameter group and can therefore be represented by their infinitesimal generators $\mathbf{X}$. For those transformations an equivalent definition is that $\mathbf{X}$ is the generator of a Lie–Bäcklund symmetry if

$$\mathbf{X}H_A = 0 \tag{24.4}$$

holds when $H_A = 0$ and all its differential consequences are taken into account.

Comparing the two definitions given above, one may wonder why we did not require that the derived system (24.2) must also be invariant under the action of $\mathbf{X}$. The reason is that the generator $\mathbf{X}$ as given by

$$\mathbf{X} = \xi^n \frac{\partial}{\partial x^n} + \eta^\alpha \frac{\partial}{\partial u^\alpha} + (D_n \eta^\alpha - u^\alpha{}_{,i} D_n \xi^i) \frac{\partial}{\partial u^\alpha{}_{,n}} + \cdots \tag{24.5}$$

and the operator D_i defined by (24.3) obey the commutation law

$$\begin{aligned}[D_k, \mathbf{X}] &= (D_k \xi^n) \frac{\partial}{\partial x^n} + (D_k \eta^\alpha - D_k \eta^\alpha + u^\alpha{}_{,i} D_k \xi^i) \frac{\partial}{\partial u^\alpha} + \cdots \\ &= (D_k \xi^n) D_n, \end{aligned} \tag{24.6}$$

so that, for example,

$$\mathbf{X} D_k H_A = 0 \tag{24.7}$$

is a consequence of $\mathbf{X}H_A = 0$ and $D_k H_A = 0$, and all the equations (24.2) are invariant provided $H_A = 0$ is invariant.

It is clear that the commutator of any pair of Lie–Bäcklund symmetries is a symmetry too.

24.2 Remarks on the structure of the set of Lie–Bäcklund symmetries

The similarity between the symmetry condition $\mathbf{X}H_A = 0$ and the differential consequence $D_k H_A = 0$ of the differential equations under consideration naturally leads to the conjecture that the operators D_k can be used to construct a symmetry generator. Indeed, it can be shown that

$$\mathbf{X}_0 = f^k(x^i, u^\alpha, u^\alpha{}_{,i}, \ldots) D_k, \tag{24.8}$$

with arbitrary functions f^k, is a symmetry generator for any set of differential equations (just as $\rho\mathbf{A}$ is always a dynamical symmetry;

compare Section 12.3). It is evident that $\mathbf{X}_0 H_A = 0$ holds as a consequence of $H_A = 0$ and $D_i H_A = 0$. What we have to prove is that $\mathbf{X}_0$ really is a generator of a Lie–Bäcklund transformation, that is, that its components obey the correct extension law. Writing $\mathbf{X}_0$ in its component representation

$$\mathbf{X}_0 = f^k \frac{\partial}{\partial x^k} + f^k u^\alpha{}_{,k} \frac{\partial}{\partial u^\alpha} + f^k u^\alpha{}_{,kj} \frac{\partial}{\partial u^\alpha{}_{,j}} + \cdots, \tag{24.9}$$

we see that we have to identify f^k with ξ^k and $f^k u^\alpha{}_{,k}$ with η^α and that because $\eta^\alpha - u^\alpha{}_{,k}\xi^k = 0$ the extension law is indeed satisfied; see Section 15.1 or Exercise 15.2.

This rather trivial class of symmetries $\mathbf{X}_0$ can be used to bring any symmetry generator $\mathbf{X}$ into its canonical form defined by

$$\hat{\mathbf{X}} = \mathbf{X} - \xi^k D_k = (\eta^\alpha - \xi^k u^\alpha{}_{,k}) \frac{\partial}{\partial u^\alpha} + \cdots \tag{24.10}$$

by simply adding an appropriate $\mathbf{X}_0$, that is, we always can assume that the symmetry under discussion has the simple form

$$\mathbf{X} = \eta^\alpha \frac{\partial}{\partial u^\alpha} + \cdots \tag{24.11}$$

(the "hat" being dropped again). For this canonical form of $\mathbf{X}$, the extension laws (15.19) and (15.20) simplify to

$$\eta^\alpha_n = \frac{\mathrm{D}}{\mathrm{D}x^n}\eta^\alpha, \qquad \eta^\alpha_{nm} = \frac{\mathrm{D}^2}{\mathrm{D}x^n\,\mathrm{D}x^m}\eta^\alpha, \ldots, \tag{24.12}$$

and in the notation (24.3) the generator $\mathbf{X}$ reads in full

$$\mathbf{X} = \eta^\alpha \frac{\partial}{\partial u^\alpha} + (D_n\eta^\alpha)\frac{\partial}{\partial u^\alpha{}_{,n}} + (D_m D_n \eta^\alpha)\frac{\partial}{\partial u^\alpha{}_{,nm}} + \cdots. \tag{24.13}$$

Although always possible, it may not always be opportune to use this canonical form, as is shown by the case of the simple translation (point transformation)

$$\mathbf{X} = \frac{\partial}{\partial x} \tag{24.14}$$

and its canonical counterpart

$$\hat{\mathbf{X}} = -u^{\alpha}{}_{,x}\frac{\partial}{\partial u^{\alpha}} + \cdots, \tag{24.15}$$

which looks like a "true" Lie–Bäcklund transformation (compare also Example 1 of Section 23.4). All generators that in their canonical form (24.11) have η^{α} that are linear in the first derivatives $u^{\alpha}{}_{,i}$, with coefficients independent of all derivatives, are disguised generators of point transformations; see equation (24.10). On the other hand, all Lie–Bäcklund generators of the form

$$\mathbf{X} = \xi^{n}(x^{i}, u, u_{,i})\frac{\partial}{\partial x^{n}} + \eta(x^{i}, u, u_{i})\frac{\partial}{\partial u} \tag{24.16}$$

are equivalent to contact transformation generators; see Exercise 24.1.

How many nontrivial Lie–Bäcklund symmetries does a given set $H_A = 0$ of differential equations admit? If we start from the canonical form $\mathbf{X} = \eta^{\alpha}\,\partial/\partial u^{\alpha}$ of the symmetry generator, we have to determine the functions η^{α} from the symmetry conditions $\mathbf{X}H_A = 0$, which are a system of linear partial differential equations for the η^{α}. The η^{α} are functions of x^n, u^{α} and all derivatives of u^{α}, and the coefficients in the symmetry conditions depend on the same set of variables. Since $\mathbf{X}H_A = 0$ need be satisfied only if $H_A = 0$ and its differential consequences (24.2) are taken into account, these variables cannot be treated as independent quantities. But if one selects *one* particular derivative (i.e., one of the highest occurring there) in each of the equations $H_A = 0$, $D_i H_A = 0, \ldots$, and uses these equations to eliminate the selected derivatives in the symmetry conditions, then one is left with a set of linear partial differential equations (in a somewhat diminished set of variables) without any additional constraints. These differential equations will *always* have solutions (if $H_A = 0$ admits solutions), which will in general depend on arbitrary functions.

As already mentioned in Section 23.1, one can give a heuristic argument indicating in what this plethora of symmetries originates. In the analogous case of the dynamical symmetries of ordinary differential equations, those symmetries are essentially translations in the space of first integrals; compare Sections 12.1 and 12.3. Their existence reflects the fact that a change in the value of the first integrals (of the initial values) maps solutions into solutions and is thus a symmetry operation. The essence of this statement is true also for partial differential equations: a change in the initial (or boundary) values maps solutions into solutions, and if we were able to express this change in terms of the solution functions and their derivatives, we would have constructed a Lie–Bäcklund symmetry.

As with the dynamical symmetries, the problem is not to decide whether Lie–Bäcklund symmetries are admitted by a given set of partial differential equations but to find practicable ways to determine and, of course, eventually exploit them. We shall devote the rest of this book to these problems.

24.3 How to find Lie–Bäcklund symmetries: some general remarks

When looking for Lie–Bäcklund symmetries, two types of Lie–Bäcklund transformations could in principle be distinguished. The first type are the Lie–Bäcklund transformations discussed so far, which do not refer to any particular differential equation. The second type results upon inserting the differential equations $H_A = 0$ under consideration into the Lie–Bäcklund transformation and its generator. When this has been done, the resulting transformations of the derivatives will in general make sense only for solutions of these differential equations since the extension laws will be satisfied only if $H_A = 0$ and its differential consequences $D_i H_A = 0$ are taken into account. Only the second type will naturally occur when determining the components of the generator $\mathbf{X}$ from the symmetry condition $\mathbf{X}H_A = 0$; as that condition need be satisfied only if $H_A = 0$ and $D_i H_A = 0, \ldots$ are taken into account, we shall always use these equations to eliminate certain derivatives and to simplify $\mathbf{X}$ if possible.

We need not bother about these types, but only keep in mind that the possibility of whether and how to use $H_A = 0$ to eliminate certain terms in the leading parts

$$\mathbf{X} = \xi^n \frac{\partial}{\partial x^n} + \eta^\alpha \frac{\partial}{\partial u^\alpha} \tag{24.17}$$

of a symmetry generator may lead to apparently different generators, generators that nevertheless describe the same transformation and symmetry. So the two generators

$$\mathbf{X} = u_{,t} \frac{\partial}{\partial u}, \qquad \mathbf{Y} = u_{,xx} \frac{\partial}{\partial u} \tag{24.18}$$

correspond to the same transformation if applied to solutions of the heat conduction equation $u_{,t} - u_{,xx} = 0$, and the generator (23.38) is trivial (zero) for solutions of the potential equation $u_{,xx} + u_{,yy} = 0$.

Finding symmetries means determining functions ξ^n, η^α (or η^α only, if we choose the gauge $\xi^n = 0$) from the symmetry conditions $\mathbf{X}H_A = 0$. In most cases, these conditions cannot be solved in full generality. Rather, one has to prescribe certain properties of the η^α to make the problem feasible, as was necessary when dealing with dynamical symmetries of ordinary differential equations. When dealing with η^α that depend only on a *finite* number of derivatives, one can, for example, prescribe the highest order of derivatives or start with η^α that are polynomials in the derivatives. To obtain symmetry generators already depending on an infinite set of derivatives in their leading part (24.15), more sophisticated techniques have to be used. But for a general system of partial differential equations no guidance on how to look and search for Lie–Bäcklund symmetries can be given.

Many examples of Lie–Bäcklund symmetries are known, and the following section will be devoted to some of them. However, to be honest, the more interesting and important symmetries have often been found in hindsight and not by solving the symmetry conditions. Rather, some generation techniques for solutions were first invented and only later recognized as originating in the existence of certain Lie–Bäcklund symmetries.

24.4 Examples of Lie–Bäcklund symmetries

1. For linear differential equations, many examples of Lie–Bäcklund symmetries are known, and in particular for linear differential equations with constant coefficients they are comparatively easy to obtain. Take, for example, the heat conduction equation. To simplify the notation, we use the abbreviations

$$u_{(0)} = u, \qquad u_{(1)} = u_{,x}, \qquad u_{(2)} = u_{,xx}, \quad \text{etc.,} \tag{24.19}$$

for the function u and its x-derivatives and write this equation as

$$H = u_{,t} - u_{,xx} = u_t - u_{(2)} = 0. \tag{24.20}$$

In the gauge $\xi = 0$, $\mathbf{X} = \eta\, \partial/\partial u$, and using the form (24.13) of the generator, the symmetry condition $\mathbf{X}H = 0$ reads

$$(D_t - D_x^2)\eta|_{H=0} = 0. \tag{24.21}$$

Since $u_{,t}$ and all its derivatives can be eliminated by means of

$$u_{,t} = u_{(2)}, \qquad u_{t(n)} = u_{(n+2)}, \qquad u_{tt} = u_{(2)t} = u_{(4)}, \quad \text{etc.,} \tag{24.22}$$

we can assume from the beginning that $\eta = \eta(x, t, u_{(i)})$ holds. If we use

$$D_x = \frac{\partial}{\partial x} + \sum_n u_{(n+1)} \frac{\partial}{\partial u_{(n)}}, \qquad D_t = \frac{\partial}{\partial t} + \sum_n u_{(n+2)} \frac{\partial}{\partial u_{(n)}}, \tag{24.23}$$

then the symmetry condition (24.21) yields

$$\left(\frac{\partial^2}{\partial x^2}+2\sum_n u_{(n+1)}\frac{\partial^2}{\partial u_{(n)}\,\partial x}\right.$$
$$\left.+\sum_{n,m} u_{(n+1)}u_{(m+1)}\frac{\partial^2}{\partial u_{(n)}\,\partial u_{(m)}}-\frac{\partial}{\partial t}\right)\eta(x,t,u_{(i)})=0 \tag{24.24}$$

[after the substitution (24.22), $H=0$ need not be taken into account any longer].

Solutions η to the symmetry condition (24.24), that is, generators of Lie–Bäcklund symmetries of the heat conduction equation, can easily be constructed. Assume, for example, that η is a function of at most second derivatives, $\eta=\eta(x,t,u,u_{(1)},u_{(2)})$. Then the coefficients of $u_{(3)}^2$ and $u_{(3)}$ in (24.24) must vanish, and this immediately gives

$$\eta = A(t)u_{(2)} + B(x,t,u,u_{(1)}), \tag{24.25}$$

where η is necessarily linear in $u_{(2)}$. Repeating this reasoning for the coefficients of $u_{(2)}^2, u_{(2)},\dots$, we see that η must be linear in all the $u_{(i)}$, with coefficients that are polynomials in x and t. We leave the details of the calculation to the reader, and state only the final result: the general solution η of the symmetry condition that depends on at most second derivatives of u is given by

$$\eta = a_1[4t^2u_{(2)}+4xtu_{(1)}+(x^2+2t)u]+a_2[2tu_{(2)}+xu_{(1)}]$$
$$+a_3u_{(2)}+a_4[2tu_{(1)}+xu]+a_5u_{(1)}+a_6u+h(x,t), \tag{24.26}$$

where the a_i are constants and h obeys $h_{,t}-h_{,xx}=0$.

In principle, we could proceed to higher order derivatives and determine the corresponding solutions η in an analogous way. But there are better methods of finding those generators that make the structure of the set of Lie–Bäcklund symmetries more transparent. We shall come back to this problem in the following section.

2. An example of a linear differential equation with nonconstant coefficients is the quantum-mechanical counterpart

$$H=-\frac{\hbar^2}{2m}(u_{,11}+u_{,22}+u_{,33})+\frac{A}{r}u+Eu=0,$$
$$r^2\equiv(x^1)^2+(x^2)^2+(x^3)^2, \qquad E<0, \tag{24.27}$$
$$\hbar^2/2m,\ A,\ E=\text{const.},$$

of the classical Kepler problem in Section 13.3. Here again one can try the *ansatz* that η depends on at most the Nth derivatives of u. Since a dependence on at most first derivatives leads to point or contact symmetries (compare Exercise 24.1), the first nontrivial case to try is $N=2$. It turns out that such Lie–Bäcklund symmetries do indeed exist and are given by the three generators

$$\mathbf{X}_{(i)} = \eta_{(i)}\frac{\partial}{\partial u},$$
$$\eta_{(i)} = \frac{x_i u}{r} - \frac{\hbar^2}{mA}u_{,i} - \frac{\hbar^2}{mA}\varepsilon_i{}^{jk}\varepsilon_{jl}{}^{n}x^l u_{,nk}. \tag{24.28}$$

These Lie–Bäcklund symmetries are the counterpart of the dynamical symmetries (13.24) corresponding to certain quadratic first integrals (Runge–Lenz vector) of the classical Kepler problem.

3. For a simple nonlinear equation such as the Burgers equation

$$u_{,t} + 2uu_{,x} - u_{,xx} = 0, \tag{24.29}$$

we can in principle repeat the above procedure and prescribe the dependence of η on the derivatives to find a symmetry. But for this particular equation we know that it is related to the heat conduction equation

$$v_{,t} - v_{,xx} = 0 \tag{24.30}$$

by means of the transformation

$$u = -v_{,x}/v = -D_x \ln v, \qquad v = \exp(-D_x^{-1}u), \tag{24.31}$$

where D_x^{-1} denotes the integration operator

$$D_x^{-1} = \int_{-\infty}^{x} \mathrm{d}x. \tag{24.32}$$

So we can use this knowledge, take a Lie–Bäcklund symmetry $\overset{(v)}{\eta}$ of the heat conduction equation and transform it to obtain a symmetry $\overset{(u)}{\eta}$ of Burgers equation. Lie–Bäcklund symmetries are invariant under transformations of variables such as (24.31); only the functional structure of their components will change. To obtain $\overset{(u)}{\eta}$, we start from the general law

$$\mathbf{X} = (\mathbf{X}u)\frac{\partial}{\partial u} + \cdots = (\mathbf{X}v)\frac{\partial}{\partial v} + \cdots \tag{24.33}$$

and compute $\overset{(u)}{\eta}$ from

$$\begin{aligned} \overset{(u)}{\eta} = (\mathbf{X}u) &= \left[(\mathbf{X}v)\frac{\partial}{\partial v} + (\mathbf{X}v_{,x})\frac{\partial}{\partial v_{,x}} + \cdots\right]u \\ &= -\left[\overset{(v)}{\eta}\frac{\partial}{\partial v} + (D_x\overset{(v)}{\eta})\frac{\partial}{\partial v_{,x}}\right]\frac{v_{,x}}{v}. \end{aligned} \tag{24.34}$$

The result is

$$\overset{(u)}{\eta} = -D_x\frac{1}{v}\overset{(v)}{\eta}, \tag{24.35}$$

where of course v must be substituted by u on the right side.

If we apply (24.35) to $\overset{(v)}{\eta} = v$, we obtain $\overset{(u)}{\eta} = 0$; a scaling $\tilde{v} = \text{const.}\, v$ does not affect u, as (24.31) clearly shows. The symmetries $\overset{(v)}{\eta} = v_{(1)}$, $\overset{(v)}{\eta} = v_{(2)}$ lead to

$$\overset{(u)}{\eta} = u_{(1)}, \qquad \overset{(u)}{\eta} = u_{(2)} - 2uu_{(1)}, \tag{24.36}$$

respectively.

4. A more complicated example of a set of nonlinear partial differential equations is provided by Einstein's vacuum field equations

$$R_{ab} = 0; \tag{24.37}$$

compare Sections 13.4 and 16.2 for notation. They are a set of 10 second order differential equations from which the 10 components g_{ab} of the metric tensor are to be determined. If we write the symmetry generator as

$$\mathbf{X} = (\mathbf{X}g_{ab})\frac{\partial}{\partial g_{ab}} = f_{ab}\frac{\partial}{\partial g_{ab}}, \tag{24.38}$$

then we have

$$\begin{aligned} \mathbf{X}g_{ab,i} &= (\mathbf{X}g_{ab})_{,i} = f_{ab,i}, \\ \mathbf{X}\Gamma^r_{st} &= \tfrac{1}{2}g^{rn}(f_{ns;t} + f_{nt;s} - f_{st;n}), \end{aligned} \tag{24.39}$$

and the symmetry condition $\mathbf{X}R_{ab} = 0$ amounts to

$$f^n{}_{a;bn} + f^n{}_{b;an} - f_{ab}{}^{;n}{}_{;n} - f^n{}_{n;ab} = 0 \qquad (\text{mod } R_{ab} = 0). \tag{24.40}$$

In the general case, no nontrivial solution to this symmetry condition has been found so far. Symmetries are known only if we restrict the solutions of $R_{ab} = 0$ to a certain set of similarity solutions by demanding that space-time admits a Killing vector field $\xi^n(x^i)$,

$$\xi_{i;n} + \xi_{n;i} = 0; \tag{24.41}$$

compare Section 18.3. The ξ^n then play the role of additional dependent variables, and (24.41) must be added to the original field equations (24.37). For a symmetry operation, (24.41) has to be invariant too.

A Lie–Bäcklund symmetry is known for the system (24.37) and (24.41). It was *not* found by solving (24.40) but by pursuing a different idea for generating new solutions from known ones. We cannot go into the details of the calculations here but shall simply present the generator to the reader to give an impression of how involved a Lie–Bäcklund symmetry may and will be. The generator reads

$$\begin{aligned} &\mathbf{X} = f_{ab}\frac{\partial}{\partial g_{ab}}, \qquad \mathbf{X}\xi^a = 0, \\ &f_{ab} = \mathbf{X}g_{ab} = -2\omega g_{ab} + 2(\xi_a\alpha_b + \xi_b\alpha_a), \end{aligned} \tag{24.42}$$

where the functions ω, α_a, and β_a (these last also being necessary later) obey

$$\begin{aligned} \omega_{,a} &= \varepsilon_{abcd}\xi^{d;c}\xi^b, \\ \alpha_{b,a} - \alpha_{a,b} &= \varepsilon_{abcd}\xi^{d;c}, \\ \beta_{b,a} - \beta_{a,b} &= 4\lambda\xi_{b;a} + 2\omega\varepsilon_{abcd}\xi^{d;c} \end{aligned} \tag{24.43}$$

and are gauged by

$$\xi^a\alpha_a = \omega, \qquad \xi^a\beta_a = \omega^2 + \lambda^2, \qquad \lambda \equiv \xi_a\xi^a \neq 0 \tag{24.44}$$

(ε_{abcd} is the totally antisymmetric Levi–Civita pseudotensor, with $\varepsilon_{1234} = -\sqrt{-g} = -\sqrt{-\|g_{ab}\|}$).

Equations (24.42) and (24.43) exhibit the nonlocal structure of the symmetry generator; to really know ω and α_a as functions of g_{ab} and ξ^a, one first has to integrate (24.43)!

24.5 Recursion operators

To find a Lie–Bäcklund symmetry, one always has to make some guess concerning the structure of its generator. A very common *ansatz* is to assume that the η^α depend on at most the Nth derivatives (and perhaps are even linear in the highest derivatives); compare the first two examples given above. A different but related approach is to search for recursion operators.

The main idea runs as follows. For any set of differential equations, the symmetry conditions $\mathbf{X}H_A = 0$ are *linear* partial differential equations for the η^α. If we have only one dependent variable u,

$$H(x^i, u, u_{,i}, u_{,ik}, \ldots) = 0,$$

$$\mathbf{X} = \eta \frac{\partial}{\partial u} + (D_i\eta)\frac{\partial}{\partial u_{,i}} + (D_iD_k\eta)\frac{\partial}{\partial u_{,ik}} + \cdots, \tag{24.45}$$

then the symmetry condition reads

$$\mathbf{F}\eta = \left(\frac{\partial H}{\partial u} + \frac{\partial H}{\partial u_{,i}} D_i + \frac{\partial H}{\partial u_{,ik}} D_iD_k + \cdots\right)\eta = 0 \qquad (\operatorname{mod} H = 0). \tag{24.46}$$

Suppose one solution η to this condition is known. Can this solution be used to generate new solutions, that is, new Lie–Bäcklund symmetries? If $\mathbf{F}\eta = 0$ was a linear differential equation with *constant* coefficients, then as well as $\eta, \eta_{,i}, \eta_{,ik}, \ldots$ would also be solutions. In generalizing this case, we may ask whether a nonconstant operator $\mathbf{L}$ exists such that for any η satisfying $\mathbf{F}\eta = 0$, $\mathbf{L}\eta, \mathbf{L}^2\eta, \ldots$ also satisfy this equation, that is,

$$\mathbf{FL}\eta = 0 \qquad (\operatorname{mod} H = 0) \tag{24.47}$$

holds. Operators $\mathbf{L}$ satisfying this condition are called recursion operators. With the help of such a recursion operator we can hope to construct infinitely many Lie–Bäcklund symmetries $\mathbf{L}^n\eta$ from, for example, a point symmetry η.

Comparing (24.47) and (24.46), we immediately see that a sufficient condition for $\mathbf{L}$ to be a recursion operator is that we can find an operator $\mathbf{M}$ such that

$$\mathbf{FL} = \mathbf{MF} \qquad (\operatorname{mod} H = 0) \tag{24.48}$$

holds.

We now shall illustrate the above ideas by a few examples that show that such recursion operators do indeed exist.

In the case of the heat conduction equation (compare the foregoing section), the operator $\mathbf{F}$ was given by

$$\mathbf{F} = D_t - D_x^2. \tag{24.49}$$

From the representation (24.23) of D_t – which is valid only for solutions of the heat conduction equations – we infer that D_t and D_x commute. But that means that $\mathbf{F}$ and D_x also commute and that D_x is a recursion operator. If one asks for the most general operators $\mathbf{L}$ and $\mathbf{M}$ which are *linear* in D_x,

$$\mathbf{L} = a_1 D_x + a_2, \qquad \mathbf{M} = a_3 D_x + a_4, \qquad a_i = a_i(x, t, u, u_x, \ldots), \tag{24.50}$$

then the operator equation (24.48), that is,

$$(D_t - D_x^2)(a_1 D_x + a_2) = (a_3 D_x + a_4)(D_t - D_x^2), \tag{24.51}$$

can easily be evaluated by comparing the coefficients of the different powers of D_x and D_t. The result is that $\mathbf{L} = \mathbf{M}$ and is a linear combination of the two recursion operators

$$\mathbf{L}_1 = D_x, \qquad \mathbf{L}_2 = 2tD_x + x \tag{24.52}$$

(a constant $\mathbf{L}$ is trivial and can be neglected). By a repeated application of these two recursion operators to the simple scaling symmetry $\eta = u$, all Lie–Bäcklund symmetries that are polynomials in the derivatives can be generated. Among them are (24.26) and also

$$\eta = u_{(k)} \tag{24.53}$$

and

$$\eta = xu_{(k)} + 2tu_{(k+1)}. \tag{24.54}$$

All these Lie–Bäcklund symmetries are closely related to the point symmetries (16.64) when the freedom in the choice of $g(x, t)$ in that equation is properly interpreted [compare Section 17.4 and equation (17.36)] and the method used in their construction rests heavily on the linearity of the heat conduction equation.

The second example is again Burgers equation. Since we know its close relation to the linear heat conduction equation, we shall not try to invent a new technique for finding recursion operators, but rather take the two recursion operators $\overset{(v)}{\mathbf{L}}_1 = D_x$ and $\overset{(v)}{\mathbf{L}}_2 = 2tD_x + x$ of the heat conduction equation and transform them to obtain the corresponding operators $\overset{(u)}{\mathbf{L}}_1$

and $\overset{(u)}{\mathbf{L}}_2$ of the Burgers equation. Since, as well as $\overset{(u)}{\eta}$ and $\overset{(v)}{\eta}$, $\overset{(u)}{\mathbf{L}}\overset{(u)}{\eta}$ and $\overset{(v)}{\mathbf{L}}\overset{(v)}{\eta}$ are (the first components of) symmetries, we can apply the transformation law (24.35) to both sets. This yields for any $\overset{(u)}{\eta}$

$$\overset{(u)}{\mathbf{L}}\overset{(u)}{\eta} = -D_x \frac{1}{v}\overset{(v)}{\mathbf{L}}\overset{(v)}{\eta} = D_x \frac{1}{v}\overset{(v)}{\mathbf{L}} v D_x^{-1}\overset{(u)}{\eta}, \tag{24.55}$$

which implies the transformation law

$$\overset{(u)}{\mathbf{L}} = D_x \frac{1}{v}\overset{(v)}{\mathbf{L}} v D_x^{-1}. \tag{24.56}$$

If we apply this law to $\mathbf{L}_1$ and $\mathbf{L}_2$, we obtain as the two fundamental recursion operators of the Burgers equation

$$\overset{(u)}{\mathbf{L}}_1 = D_x - u - u_{,x} D_x^{-1},$$

$$\overset{(u)}{\mathbf{L}}_2 = 2tD_x + x - 2tu + (1 - 2tu_{,x}) D_x^{-1}. \tag{24.57}$$

What may be a surprise is the explicit occurrence of the integration operator D_x^{-1}, which leads to a simple although perhaps unexpected structure of the Lie–Bäcklund symmetries.

The idea of looking for recursion operators when trying to find Lie–Bäcklund symmetries can be applied also to arbitrary evolution equations, that is, equations of the form

$$u_{,t} + f(u, u_{(1)}, \ldots, u_{(N)}) = 0, \tag{24.58}$$

where $u_{(n)}$ denotes the nth derivative of u with respect to x (the heat conduction and Burgers equations are special simple cases of evolution equations). As x and t do not explicitly occur in f, those evolution equations always admit two point symmetries $\mathbf{X}_1 = \partial/\partial x$ and $\mathbf{X}_2 = \partial/\partial t$, which in the gauge $\mathbf{X} = \eta\, \partial/\partial u$ (and using $u_{,t} = -f$) appear as

$$\mathbf{X}_1 = -u_{(1)}\frac{\partial}{\partial u}, \qquad \mathbf{X}_2 = f(u, \ldots, u_{(N)})\frac{\partial}{\partial u}. \tag{24.59}$$

In the case of the Burgers equation, these two generators are related by the recursion operator $\overset{(u)}{\mathbf{L}}_1$ since here $f = 2uu_{,x} - u_{,xx}$ and $-\overset{(u)}{\mathbf{L}}_1 u_{(1)} =$

$-\overset{(u)}{\mathbf{L}}_1 u_{,x} = -u_{,xx} + 2uu_{,x}$ hold. So it may be worth testing in other cases whether a recursion operator $\mathbf{L}$ satisfying

$$\mathbf{L}u_{(1)} = -f(u, \ldots, u_{(N)}) \tag{24.60}$$

exists. Or, alternatively, one may try to find a new (third) Lie–Bäcklund symmetry $\eta = \eta(u, \ldots, u_{(2N-1)})$ – corresponding to $\mathbf{L}f$ – which should be linear in $u_{(2N-1)}$ if f is linear in $u_{(N)}$.

24.6 Exercises

1 Show that a generator $\mathbf{X}$ of the form (24.16) is equivalent to a generator $\mathbf{Y}$ of a contact transformation (21.10) and (21.11) with generating function $\Omega = \eta - u_{,n}\xi^n$ in the sense that $\mathbf{X} - \mathbf{Y}$ is a suitably chosen $\mathbf{X}_0$ of the form (24.8).

2 Let $\mathbf{X}$ be a Lie–Bäcklund symmetry. When is $\lambda\mathbf{X}$ a symmetry too?

3 Show that η as given by (24.26) is the general Lie–Bäcklund symmetry of the heat conduction equation that depends on at most second derivatives. Which of these symmetries are point or contact symmetries? (Note that the heat conduction equation can be used to change the form of η.)

4 Show that for any vector field ξ^n the generator $\mathbf{X} = \mathscr{L}_\xi g_{ab}\,\partial/\partial g_{ab}$ is a generator of a Lie–Bäcklund symmetry, that is, that $\mathscr{L}_\xi g_{ab} = \xi_{a;b} + \xi_{b;a}$ is a solution of (24.40). Why is this symmetry trivial?

5 Show that there are no Lie–Bäcklund symmetries $\mathbf{X} = \eta\,\partial/\partial u$ of $u_{,xy} = \sin u$ with η containing derivatives of at most second order.

6 Find the general recursion operator $\mathbf{L} = b_1 D_t + b_2 D_x + b_3$ of the heat conduction equation and show that it can be constructed from $\mathbf{L}_1$ and $\mathbf{L}_2$.

7 Find a recursion operator $\mathbf{L}$ for the Korteweg–deVries equation $u_{,xxx} + 6uu_{,x} + u_{,t} = 0$ by starting from (24.60). Note that it must be shown that $\tilde{\eta} = \mathbf{L}\eta$ is a symmetry for *any* symmetry η.

25

How to use Lie–Bäcklund symmetries

Simple convincing applications of Lie–Bäcklund symmetries are rare. This mirrors the fact that most of the known Lie–Bäcklund symmetries either are rather trivial (as those of linear partial differential equations with constant coefficients are) or have been found in hindsight, emanating from very complex techniques for generating solutions of partial differential equations. Because of this complexity of the more interesting examples, we want to discuss here only rather briefly a few lines of thought.

25.1 Generating solutions by finite symmetry transformations

If a special solution of a partial differential equation is known and a finite Lie–Bäcklund symmetry is at hand, this symmetry could provide us with a one-parameter set of solutions generated from the original one – in that respect there is no difference from a Lie point symmetry. But whereas in the case of point symmetries we could be rather sure that – given the infinitesimal generator – we could construct the finite transformation, the situation is different here: in most cases it will be impossible to integrate the system

$$\frac{\mathrm{d}\tilde{u}^\alpha}{\mathrm{d}\varepsilon} = \eta^\alpha(x^i, \tilde{u}^\beta, \tilde{u}^\beta{}_{,i}, \dots) \tag{25.1}$$

in a straightforward way; compare Section 23.3 and Example 2 in Section 23.4.

In some cases, a way out is to attack the problem indirectly. Constructing the finite transormation is essentially the same as transforming the generator to its normal form $\mathbf{X} = \partial/\partial\psi$. The idea now is to divide this task into two steps: first to transform the generator of the Lie–Bäcklund symmetry into that of a Lie point symmetry by introducing new variables ("potentials") and then to exponentiate this infinitesimal point transformation. Finding these new variables is, in principle, of equal complexity to solving (25.1), but it may happen that the symmetry generator η^α already indicates which variables one should take.

Consider, for example, the Lie–Bäcklund symmetry

$$\mathbf{X} = [-2\omega g_{ab} + 2(\xi_a\alpha_b + \xi_b\alpha_a)]\frac{\partial}{\partial g_{ab}} \tag{25.2}$$

of Einstein's vacuum field equation; compare Example 4 in Section 24.4. The original dependent variables are the metric tensor components g_{ab}, but in the formulation of the symmetry the quantities ξ^a, ω, α_a, β_a, and λ quite naturally appear. To see whether candidates for new variables are among them [variables that make (25.2) a point symmetry], we determine how $\mathbf{X}$ acts on them by applying $\mathbf{X}$ to the defining relations (24.43) and (24.44). We cannot give the details of the calculations here, but simply list the results, which are

$$\mathbf{X}\omega = \omega^2 - \lambda^2, \qquad \mathbf{X}\lambda = 2\omega\lambda, \tag{25.3}$$

$$\mathbf{X}\alpha_a = 2\omega\alpha_a - \beta_a, \qquad \mathbf{X}\beta_a = 2(\omega^2 + \lambda^2)\alpha_a, \tag{25.4}$$

and

$$\mathbf{X}\xi_a = 2\lambda\alpha_a, \qquad \mathbf{X}h_{ab} \equiv \mathbf{X}(g_{ab} - \xi_a\xi_b/\lambda) = -2\omega h_{ab}. \tag{25.5}$$

Since no derivatives of the variables appear on the right side of equations (25.3)–(25.5), the operator $\mathbf{X}$ *is* already the generator of a point symmetry in these variables!

One can now proceed to exponentiate the infinitesimal generator, that is, to solve (25.1) for these new variables. This is rather easy for the variables λ and ω but more involved for the rest. Introducing the complex Ernst potential $E = \lambda + i\omega$ into (25.3) leads to

$$\mathbf{X}E = \mathbf{X}(\lambda + i\omega) = -iE^2, \tag{25.6}$$

and this is exponentiated by $\tilde{E} = E/(1 + i\varepsilon E)$ or

$$\begin{aligned} \tilde{\lambda} &= \lambda/[1 - 2\varepsilon\omega + \varepsilon^2(\lambda^2 + \omega^2)], \\ \tilde{\omega} &= \tilde{\lambda}[\omega - \varepsilon(\omega^2 + \lambda^2)]/\lambda. \end{aligned} \tag{25.7}$$

The other equations finally lead to

$$\begin{aligned} \tilde{g}_{ab} &= (\lambda g_{ab} - \xi_a \xi_b)/\tilde{\lambda} + \tilde{\xi}_a \tilde{\xi}_b/\tilde{\lambda}, \\ \tilde{\xi}_a &= \xi_a + \tilde{\lambda}(2\varepsilon\alpha_a - \varepsilon^2 \beta_a). \end{aligned} \tag{25.8}$$

Equations (25.7) and (25.8) describe how a finite Lie–Bäcklund symmetry with generator (25.2) acts on a solution g_{ab} of Einstein's field equations. To really calculate the new solution $\tilde{g}_{ab}$, one needs to know not only the old solution g_{ab}, but also the functions ω, λ, α_a, and β_b that are to be determined from the old solution via their defining equations (24.43) and (24.44).

This example was the starting point for introducing generation methods for solving Einstein's field equations. Although Lie–Bäcklund symmetries were not mentioned there, they were practically always exploited.

25.2 Similarity solutions for Lie–Bäcklund symmetries

In Chapter 18 we discussed at some length how Lie point symmetries can be used to introduce similarity variables in order to reduce the number of independent variables and thus gain access to special classes of solutions. For a given symmetry **X**, the similarity variables φ^{Ω} were defined to obey $\mathbf{X}\varphi^{\Omega} = 0$, and similarity solutions $u^{\alpha} = u^{\alpha}(x^n)$ had to be invariant under the action of **X** and to satisfy

$$\mathbf{X}[u^{\alpha} - u^{\alpha}(x^n)] = \eta^{\alpha} - \xi^n u^{\alpha}{}_{,n} = 0 \tag{25.9}$$

in addition, of course, to the given system $H_A = 0$ of differential equations; compare Section 18.2.

It is easy to generalize this concept to Lie–Bäcklund symmetries: we have to demand that (25.9) is valid here also! If we prefer to write the symmetry generator **X** in its canonical form

$$\mathbf{X} = \eta^{\alpha}(x^n, u^{\beta}, u^{\beta}{}_{,n}, \ldots)\frac{\partial}{\partial u^{\alpha}}, \tag{25.10}$$

then similarity solutions of this Lie–Bäcklund symmetry have to satisfy

$$\eta^{\alpha}(x^n, u^{\beta}, u^{\beta}{}_{,n}, \ldots) = 0. \tag{25.11}$$

Because of the relation (24.10) between the general and the canonical form of a Lie–Bäcklund symmetry, it is obvious that point symmetries are properly included in this definition and the same is true for contact symmetries, where $\eta = -\Omega$ was the generating function; compare Section 21.3.

To prove that the two sets $H_A = 0$ and $\eta^\alpha = 0$ do not contradict each other and thus can indeed possess a common system of solutions, we imagine that the Lie–Bäcklund symmetry has been transformed to its normal form $\mathbf{X} = \partial/\partial\psi$, where ψ is a (new) independent variable. For this generator, the canonical form is $\mathbf{X} = -u^\alpha{}_{,\psi}\,\partial/\partial u^\alpha$, and $\eta^\alpha = 0$ amounts to $u^\alpha{}_{,\psi} = 0$; similarity solutions must be independent of ψ. Since we know from $\mathbf{X}H_A = 0$ that the H_A are (or can be rewritten such that they are) independent of ψ, solutions independent of ψ can exist; $H_A = 0$ and $\eta^\alpha = 0$ do not contradict each other.

For two simple examples, we again take the heat conduction equation

$$H = u_{,t} - u_{,xx} = 0 \tag{25.12}$$

and its Lie–Bäcklund symmetries discussed in Section 24.4.

The symmetries

$$\eta = u_{(k)} = \left(\frac{\mathrm{D}}{\mathrm{D}x}\right)^k u \tag{25.13}$$

give similarity solutions that are polynomials in x with (in general) time-dependent coefficients. So for $k = 1$ the common solution of $H = 0$ and $u_{,x} = 0$ is $u = \text{const.}$, and for $k = 3$ we infer from $u_{,t} - u_{,xx} = 0$ and $u_{,xxx} = 0$ that

$$u = a_0 t + a_1 + \tfrac{1}{2}a_0 x^2, \qquad a_i = \text{const.}, \tag{25.14}$$

holds.

The member $k = 2$ of the Lie–Bäcklund symmetries (24.54) leads to the condition

$$xu_{,xx} + 2tu_{,xxx} = 0. \tag{25.15}$$

It can be integrated once with respect to x,

$$u_{,xx} = f(t)\exp(-x^2/4t). \tag{25.16}$$

The condition $u_{,xx} = u_{,t}$, that is, $u_{,xxt} = u_{,xxxx}$, can be satisfied only if $f = 1/\sqrt{t}$, and the final result is that

$$u = a_1 \int^t \exp\left(-\frac{x^2}{4t}\right)\frac{\mathrm{d}t}{\sqrt{t}} + a_2 x + a_3, \qquad a_i = \text{const.}, \tag{25.17}$$

are the similarity solutions corresponding to the generator $\mathbf{X} = (xu_{,xx} + 2tu_{,xxx})\,\partial/\partial u$.

It is clear that in the case of such a simple differential equation as the heat conduction equation the practical value of this plethora of similarity solutions (corresponding to the infinitely many Lie–Bäcklund symmetries) is small since many other ways of obtaining closed-form solutions are known.

25.3 Lie–Bäcklund symmetries and conservation laws

For ordinary differential equations there was an intimate connection between symmetries and first integrals, that is, functions $\varphi = \varphi(x, y, y', \dots)$ that are constant,

$$\frac{\mathrm{d}}{\mathrm{d}x}\varphi \equiv D_x\varphi = 0, \tag{25.18}$$

if the differential equation was taken into account. In particular, when the differential equations were derivable from a Lagrangian, the first integrals belonging to a given Noether symmetry could easily be constructed; compare Section 10.3. Since those first integrals could be used to lower the order of the differential equations, their existence and construction were an essential part of the integration procedure.

For partial differential equations, the analogue of first integrals are conserved quantities $\varphi^n = \varphi^n(x^i, u^\alpha, u^\alpha{}_{,i}, \dots)$ that obey conservation laws

$$\frac{\mathrm{D}}{\mathrm{D}x^n}\varphi^n \equiv D_n\varphi^n = 0 \tag{25.19}$$

if the differential equations $H_A = 0$ are taken into account. There is again a connection between symmetries and conservation laws for systems derivable from a Lagrangian, a connection governed by Noether's theorems. Although those conservation laws and their derivation from symmetries of the action principle are widely discussed in the literature and play an important role in the understanding and interpretation of, for example, the physical theory underlying the set of partial differential equations, they are in general not of much help for the actual integration of the differential equations. For that reason we do not intend to discuss this subject in detail.

We only want to mention an important property of Lie–Bäcklund symmetry generators $\mathbf{X}$ that can be used to generate conserved quantities. If the generators are taken in their canonical form

$$\mathbf{X} = \eta^\alpha \frac{\partial}{\partial u^\alpha} + \cdots, \tag{25.20}$$

then because of $\xi^n = 0$, equation (24.6) implies that $\mathbf{X}$ and $D_n = \mathrm{D}/\mathrm{D}x^n$ commute,

$$[\mathbf{X}, D_n] = 0. \tag{25.21}$$

So if there is a conserved quantity φ^n satisfying $D_n\varphi^n = 0$, then we can apply $\mathbf{X}$ to this equation and obtain in view of (25.21) the relation

$$D_n(\mathbf{X}\varphi^n) = 0. \tag{25.22}$$

This shows that $\psi^n = \mathbf{X}\varphi^n$ – if not zero – is a possibly new conserved quantity. In cases where (e.g., due to the existence of a recursion operator) infinitely many Lie–Bäcklund generators $\mathbf{X}$ are known, one usually also has to hand infinitely many conservation laws. Conversely, the existence of many conservation laws may indicate the possible existence of a recursion operator and thus the integrability of the differential equation.

To give a simple example, we take the heat conduction equation $u_{,t} - u_{,xx} = 0$. Here $\varphi^n = (u, -u_{,x})$ is a (rather trivial) conserved quantity since

$$D_n\varphi^n = D_t u + D_x(-u_{,x}) = u_{,t} - u_{,xx} = 0 \tag{25.23}$$

holds. Application of the generators (24.54), that is, of

$$\mathbf{X} = [xu_{(k)} + 2tu_{(k+1)}]\frac{\partial}{\partial u} + \cdots, \qquad u_{(k)} \equiv \left(\frac{\mathrm{D}}{\mathrm{D}x}\right)^k u, \tag{25.24}$$

to this φ^n yields a whole series

$$\psi^n = (xu_{(k)} + 2tu_{(k+1)}, -u_{(k)} - xu_{(k+1)} - 2tu_{(k+2)}) \tag{25.25}$$

of new conserved quantities (derivatives with respect to t have been eliminated by means of $u_{,t} = u_{,xx}$).

25.4 Lie–Bäcklund symmetries and generation methods

In the last two decades, powerful methods have been developed for solving special classes of partial differential equations, for example, evolution equations such as (24.58) and other equations (describing solitons) in two independent variables. These methods use techniques such as Bäcklund transformations, inverse scattering, Lax pairs, prolongation structures, and so on, and they often result in rules for finding (infinitely many) new solutions if one – often trivial – solution is known: they are generation methods.

Generating a new solution from an old one means performing a mapping in the space of solutions. Since by definition any mapping in the space of solutions is a Lie–Bäcklund symmetry, it is clear that there must be a close connection between these generation techniques and the exploitation of (some of) the always present Lie–Bäcklund symmetries. In particular, one might guess that many of the proposals for linearizing a given system of differential equations – attempts that typify some of these techniques – correspond to attempts to make a particular Lie–Bäcklund symmetry a point symmetry and that the "hidden" symmetries sometimes referred to in the context of these generation methods are Lie–Bäcklund symmetries.

But although in principle the existence of a close correspondence between generation techniques and Lie–Bäcklund symmetries is clear, most of the details of this correspondence are by no means obvious, are not yet understood, and have not yet been thoroughly examined. This is partially due to the fact that Lie–Bäcklund symmetries are not so widely known. But the deeper reason certainly is that the mere knowledge about these symmetries is not of much help. One has to develop methods to exploit them, different methods for different types of partial differential equations, and this is exactly what was done by inventing the generation techniques. The only aspect that may make the Lie–Bäcklund symmetry point of view superior is that it offers a unifying view of all these diverse methods and may thus perhaps lead to new generation techniques. But this is a task for future research.

25.5 Exercises

1 Find the differential equation of the similarity solutions to the Burgers equation (24.29) corresponding to the generators $\eta = \mathbf{L}_1 u_{,x} = (D_x - u - u_{,x} D_x^{-1}) u_{,x}$ and $\eta = (\mathbf{L}_1)^2 u_{,x}$.

2 Write the Korteweg–deVries equation (16.72) in the form $D_i \varphi^i = 0$ of a conservation law and derive further conservation laws by application of the Lie point symmetries (16.73) and of the symmetry with generator $\mathbf{X} = (D_x^2 + 4u + 2u_{,x} D_x^{-1})^2 u_{,x}\, \partial/\partial u + \cdots$.

3 What is the analogue of (25.21) in the case of ordinary differential equations?

Appendix A

A short guide to the literature

Ordinary differential equations

E. Kamke, *Differentialgleichungen – Lösungsmethoden und Lösungen Bd. 1.* Akademische Verlagsgesellschaft Geest u. Portig, Leipzig, 1951. (Reprinted, Chelsea, New York, 1959.)

E. L. Ince, *Ordinary differential equations.* Longmans, Green and Co., London, 1926. (Reprinted, Dover, New York, 1956.)

G. M. Murphy, *Ordinary differential equations and their solutions.* Van Nostrand, Princeton, 1960.

Group theory

L. P. Eisenhart, *Continuous groups of transformations.* Dover, New York, 1961.

R. Gilmore, *Lie groups, Lie algebras and some of their applications.* Wiley, New York, 1974.

B. G. Weybourne, *Classical groups for physicists.* Wiley, New York, 1974.

Symmetries of differential equations in general

S. Lie, *Vorlesungen über Differentialgleichungen mit bekannten infinitesimalen Transformationen.* Teubner, Leipzig, 1912.

L. V. Ovsiannikov, *Group analysis of differential equations.* Academic, New York, 1982.

N. H. Ibragimov, *Transformation groups applied to mathematical physics.* Reidel, Boston, 1985.

P. J. Olver, *Application of Lie groups to differential equations* (Graduate Texts in Mathematics 107). Springer, New York, 1986.

Contact transformations

S. Lie, F. Engel, *Theorie der Transformationsgruppen II*. Teubner, Leipzig, 1890.

Noether and Cartan symmetries

W. Sarlet, F. Cantrijn, Generalizations of Noether's theorem in classical mechanics, *SIAM Review 23*, 467 (1981).

Separation of variables

W. Miller, Jr., *Symmetry and separation of variables*. Addison-Wesley, Reading, Mass., 1977.
V. I. Arnold. Mathematical methods of classical mechanics. Springer, New York, 1978.

Lie–Bäcklund symmetries

R. L. Anderson, N. H. Ibragimov, *Lie–Bäcklund transformations in applications*, SIAM Studies in Appl. Math. Vol. 1. SIAM, Philadelphia, 1979.
A. M. Vinogradov, Local symmetries and conservation laws, *Acta Applicandae Math. 2*, 21 (1984).
I. S. Krasilshchik, A. M. Vinogradov, Nonlocal symmetries and the theory of coverings, *Acta Applicandae Math. 2*, 79 (1984).
A. S. Fokas. A symmetry approach to exactly solvable evolution equations, *J.M.P. 21*, 1318 (1980).

Generation methods

M. J. Ablowitz, H. Segur, *Solitons and the inverse scattering transform*. SIAM, Philadelphia, 1981.
F. Calogero, A. Degasperis, *Spectral transform and solitons*. North-Holland, Amsterdam, 1982.
C. Hoenselaers, W. Dietz (Eds.), *Solutions of Einstein's equations: techniques and results* (Lecture Notes in Physics 205). Springer, Berlin, 1984.

Appendix B

Solutions to some of the more difficult exercises

2.2 An arbitrary function $\psi(x, y)$ satisfying $\mathbf{X}\psi = 0$ can be added to s; for the example, ψ is a function of x/y.

2.3 See equation (4.26).

2.4 No.

3.1 Differentiate y n times and determine the a^α from the resulting system.

3.2 Only if μ is independent of the derivatives $y^{(i)}$.

4.2 The general symmetry of $y'' + y = 0$ is given by $\mathbf{X} = [a_1 y \sin(x + a_2) + a_7 \sin(2x + a_8) + a_6]\partial/\partial x + [a_1 y^2 \cos(x + a_2) + y\{a_7 \sin(2x + a_8) + a_5\} + a_3 \sin(x + a_4)]\partial/\partial y$.

4.3 Since the general solution has the form $y = c_1 y_1(x) + c_2 y_2(x)$, the transformation $\tilde{y} = y/y_2$, $\tilde{x} = y_1/y_2$ will give the desired form $\tilde{y} = c_1 \tilde{x} + c_2$ of the solution and lead to $\mathrm{d}^2\tilde{y}/\mathrm{d}\tilde{x}^2 = 0$.

4.4 For a general position of the P_i, the eight equations $a^N \xi_N = 0 = a^N \eta_N$ will not have a (nontrivial) solution a^N.

4.6 Write the solution as $\varphi = f(x, y) = \text{const.}$ and take φ as the new dependent variable.

5.2 The first steps are $\mathbf{X} = \partial/\partial t$, $ds/dt = -\tan(t + \varphi_0)$.

5.3 Compare Exercise 2.2.

5.4 *One* reduction is simple: use, e.g., $\mathbf{X} = \partial/\partial x$ and write $y' = p(y)$. Further reductions are possible but need more elaborate techniques; compare the following chapters.

6.1 Because of the special structure of (4.26), it suffices to check the commutators of those generators that could give ξ and η nonlinear in x and y.

6.2 $\mathbf{X}_1 = \partial/\partial x$; $\mathbf{X}_2 = x\,\partial/\partial x$; $\mathbf{X}_3 = x^2\,\partial/\partial x$.

6.3 If as well as $\mathbf{A}\varphi^\alpha = 0$, $\mathbf{X}_N\varphi^\alpha = 0$ was also true for all α $(\alpha = 1, \ldots, n)$, then $\mathbf{A} = \lambda\mathbf{X}_N$ must hold; compare Section 3.2. But this is impossible because of the very structure of $\mathbf{A}$ and $\mathbf{X}_N$.

6.8 The Wronskian of the solutions $u_k(x)$ is different from zero.

7.4 See Table 7.1 and Exercise 4.3.

7.5 The generators commute, and we have $\Delta = -2$. Equations (7.39) and (7.44) yield $\varphi = e^x(y - y' + x^2 - 2x + 2)/2$, $\psi = -e^{-x}(y + y' + x^2 + 2x + 2)/2$, and $y = \varphi e^{-x} - \psi e^x - 2 - x^2$.

8.1 $\mathbf{X}_1$ and $\hat{\mathbf{X}}_2 = \mathbf{X}_2 + \mathbf{X}_3$ satisfy $[\mathbf{X}_1, \hat{\mathbf{X}}_2] = -\hat{\mathbf{X}}_2$.

8.2 Equation (8.3) provides only the first integral $\varphi = y'/(xy + 3y/2)$. Another first integral is, e.g., $\psi = y'y^{-5/3}$, and the solution is $y = a(1 + bx)^{-3/2}$.

8.3 Equation (8.10), with $(\varphi, \psi) \rightarrow (x, y)$, gives $\mathbf{X}_1 = y\,\partial/\partial x - x\,\partial/\partial y$, $\mathbf{X}_2 = -xy\,\partial/\partial x + (1 + y^2)\,\partial/\partial y$, and $\mathbf{X}_3 = (1 + x^2)\,\partial/\partial x + xy\,\partial/\partial y$.

8.4 Substituting $y = s(x)x$, one sees that the group is the group of rotations, and the differential equation is $s'' = 0$. Equation (8.3) can be used to find the solutions.

9.1 The (abelian) G_n is provided by (4.38).

9.4 The obvious symmetries are $\partial/\partial x$, $x\,\partial/\partial x$, and $y\,\partial/\partial y$. The differential equation can be written as $(y^{-1/2})''' = 0$.

9.6 The solution is $x = c_1 + c_2 e^{-at}(a\cos t - \sin t)$, $y = c_3 + c_2 e^{-at}(a \sin t + \cos t)$.

10.1 Of the Lie symmetries $\xi = M_n(t)q^n + N$, $\eta^a = q^a \dot{M}_n q^n + B^a{}_n q^n + C^a(t)$, with $m\ddot{M}_n + KM_n = 0$, $m\ddot{C}^a + KC^a = 0$, only those with $M_n = 0$, $B_{an} = -B_{na}$ are Noether symmetries. The first integral related to $\xi = 0$, $\eta^a = C^a$ is $\varphi = \dot{C}_a q^a - C_a \dot{q}^a$.

10.2 No.

11.5 For $n = 0$: $f_{,yy} = 0$; for $n = 1$: $f_{,y} = 0$; for $n \geqslant 2$: $f = 0$.

12.2 $\rho = y'\mathring{\eta}$, $\varphi = -y^2 - y'^2$.

13.2 For three differential equations of second order, the maximum number of independent first integrals is six. Six first integrals independent of t are functionally dependent: we have $\phi H + \Psi^n Y_n + \phi^n X_n = 0$.

14.1 If they commute.

16.2 $\mathbf{X} = B(\zeta, w)\partial/\partial\zeta + F(\zeta, w)\partial/\partial w$ + complex conjugate.

17.2 No.

17.3 $\tilde{u} = u + \varepsilon$, $\tilde{t} = t + a_4\varepsilon$, $\tilde{x} = x + (a_1 + 6t)\varepsilon + 3a_4\varepsilon^2$,

$u = \varepsilon - 2/[x - (a_1 + 6t)\varepsilon + 3a_4\varepsilon^2]^2$.

18.1 $\varphi_1 = u$, $\varphi_2 = x/y$, $\varphi_3 = w = v - a\ln y$, $uu' + \varphi^2 u' = -a$, $w = -\int u'\varphi \, d\varphi$.

18.2 Yes.

18.3 $\varphi = t$, $w = u - x/6t$, $w + \varphi w' = 0$.

18.5 $\sigma = -h_{,t}/h_{,x}$.

19.1 $I_1 = Px^{-3/2} = w$; $I_2 = x^2(w^2{}_{,x} + w^2{}_{,y}) = w^2{}_{,s}$ if $s = \ln z$; the *ansatz* $w_{,s} = f(w)$ amounts to $I_2 = f(I_1)$.

19.2 $\varphi_1 = u$, $\varphi_2 = \varphi = a_2(x^2 - t^2)/2 + a_3 x - a_1 t$. The differential equation $u_{,\varphi\varphi}(2a_2\varphi + a_3^2 - a_1^2) + 2a_2 u_{,\varphi} = \sin u$ inherits $\mathbf{X} = \partial/\partial\varphi$ if $a_2 = 0$.

20.2 Yes.

22.3 $\alpha_1 = du \wedge dx + dv \wedge dy$, $\alpha_2 = u\,du \wedge dy - dv \wedge dx$.

22.4 $x_{,uu} + x_{,vv} = x_{,w}$.

23.2 No.

24.2 If $D_n \lambda = 0$.

24.7 $\mathbf{L} = -D_x^2 - 4u - 2u_{,x} D_x^{-1}$.

25.1 $u_{,t} = 0$, $u_{,x} - u^2 = \text{const.}$; $u_{,t} = u_{,xx} - 2uu_{,x}$, $u_{,xx} - 3uu_{,x} + u^3 = \text{const.}$

25.2 $[\mathbf{X}, \mathbf{A}] = 0$.

Index

Printed in the United States
By Bookmasters